AF307143

Rainer Bäuerle

Fully screen printed PTC based sensor array and OFET characterization

disserta
Verlag

Bäuerle, Rainer: Fully screen printed PTC based sensor array and OFET characterization, Hamburg, disserta Verlag, 2024

Buch-ISBN: 978-3-95935-632-9
PDF-eBook-ISBN: 978-3-95935-633-6
Druck/Herstellung: disserta Verlag, Hamburg, 2024
Cover: © disserta Verlag

Bibliografische Information der Deutschen Nationalbibliothek:
Die Deutsche Nationalbibliothek verzeichnet diese Publikation in der Deutschen Nationalbibliografie; detaillierte bibliografische Daten sind im Internet über http://dnb.d-nb.de abrufbar.

© disserta Verlag, Imprint der Bedey & Thoms Media GmbH
Hermannstal 119k, 22119 Hamburg
http://www.disserta-verlag.de, Hamburg 2024
Printed in Germany

Fully screen printed PTC based sensor array
and
OFET characterization

Von der Fakultät für Elektrotechnik, Informationstechnik, Physik
der Technischen Universität Carolo-Wilhelmina zu Braunschweig

zur Erlangung des Grades eines Doktors

der Ingenieurwissenschaften (Dr.-Ing.)

genehmigte Dissertation

von Rainer Bäuerle

aus Wertheim

eingereicht am: 15.02.2024

mündliche Prüfung am: 07.05.2024

1. Referent: Prof. Dr.-Ing. habil. Wolfgang Kowalsky
2. Referent: Prof. Dr. Erwin Peiner

Druckjahr: 2024

Dissertation an der Technischen Universität Braunschweig,
Fakultät für Elektrotechnik, Informationstechnik, Physik

Fully screen printed PTC based sensor array and OFET characterization Sensors are very important devices in today's world with Industry 4.0, large-scale automation, and the rise of artificial intelligence which requires a lot of information to train and work properly. In this work, a new sensor concept is introduced and investigated for two applications, liquid level and surface airflow sensing. The fully screen printed sensor consists of an array of positive temperature coefficient heaters and shunt resistors. Resistance of the self-heated material changes depending on the outgoing heat flux and is recorded by readout electronics to obtain spatial resolution of heat flux. To demonstrate its capabilities, the sensor is applied to the outside of a polypropylene cup. Air, water, and oil have different thermal properties and are distinguishable by the sensor, allowing the level of liquids to be determined. An alternative application of the concept is an airflow sensor. Printed as a 4×4 matrix, it can map surface airflow over the sensor. Alternative sensor matrices are often integrated into an active matrix with (organic) field effect transistors switching between pixels. The BASF material XPRD01B06 is intended for such OFETs and is characterized in this work. Batch-to-batch variations are sufficiently low. OFETs of the material show strong temperature dependence in the beginning. However, after optimization of fabrication the temperature dependence is minimized. The sensor concept and applications introduced offer alternatives for existing devices and the OFETs are already successfully integrated in active sensor matrices.

Vollständig siebgedrucktes, auf Kaltleiter basierendes Sensorarray und Charakterisierung von OFETs Sensoren sind heute sehr wichtige Bauteile, bedingt durch Industrie 4.0, umfassender Automation und dem Aufkommen künstlicher Intelligenz, welche viele Informationen zum Trainieren und Arbeiten benötigt. In dieser Arbeit wird ein neues Sensorkonzept vorgestellt und in zwei Anwendungen, als Flüssigkeitsfüllstands- und Oberflächenluftstromsensor, getestet. Der vollständig durch Siebdruck hergestellte Sensor besteht aus einem Array aus Kaltleitern und Messwiderständen. Der Widerstand des selbstheizenden Kaltleiters ändert sich abhängig vom Wärmefluss und wird ausgelesen, um eine flächenmäßige Darstellung des Wärmeflusses zu erhalten. Um die Fähigkeiten des Sensors zu veranschaulichen, wird der Sensor außen auf ein Polypropylenbecher angebracht. Luft, Wasser und Öl haben unterschiedliche thermische Eigenschaften und werden vom Sensor unterschieden, sodass der Füllstand der jeweiligen Flüssigkeit gemessen wird. Eine alternative Anwendung des Konzepts ist ein Luftstromsensor. Als 4×4 Matrix erfasst der Sensor den Oberflächenluftstrom. Andere Sensormatrizen nutzen oft eine aktive Matrix, in der (organische) Feldeffekttransistoren zwischen den einzelnen Pixeln schalten. XPRD01B06, ein Material von BASF, ist für solche Transistoren angedacht und in dieser Arbeit charakterisiert. Die Streuung zwischen Chargen ist gering. OFETs aus XPRD01B06 zeigen zu Beginn der Arbeit eine starke Abhängigkeit von der Temperatur. Nach Optimierung der Herstellung jedoch ist die Temperaturabhängigkeit ausreichend verringert. Das vorgestellte Sensorkonzept und seine Anwendungen bieten Alternativen für bereits vorhandene Instrumente und die OFETs sind bereits erfolgreich in aktiven Sensormatrizen eingefügt.

Contents

Contents

1. Introduction

Sensors are part of every device nowadays. Cars incorporate sensors to obtain details about their environment to a degree that allows autonomous driving [1, 2], people wear smartwatches with over 10 different kinds of sensors [3], and smart clothes with multiple sensors have emerged [4, 5]. Research and development of sensors never stopped and is driven by the desire to gain further and more detailed information on everything. Discovery of electrically conductive polymers and subsequent research in conductive and semiconductive organic materials opened new and vast possibilities for many sensors [6–9]. As the discovery revolutionized thinking about plastics in electricity, the discoverers were awarded the Nobel Prize in 2000 [10]. Since then, a vast selection of different organic materials has been found and tailored to suit different applications, many of which are of the sensory type. Materials change their properties, such as conductivity, depending on outer stimuli, meaning they can sense aspects of their environment. The new group of materials not only allows for new sensors by material properties, but also by their processibility. While most classical inorganic (semi-)conductors require high vacuum, high temperatures and further costly process conditions, many organic materials can be solution-processed, thereby circumventing the cost intensive requirements. Devices made out of organic electronics can thus be fabricated and offered at a much cheaper price. Methods for solution-processing include, among others, spin-coating, blade-coating, and, most important for structured devices, various printing techniques [11–15]. Inkjet, aerosol, roll-to-roll, and screen printing can be utilized, depending on the materials' properties and ability to form suitable inks. These printing options allow for cost-effective mass production, production runs with low quantities to single prototypes, or individual customization. Because most polymers are flexible and forgiving to bends, devices can be fabricated on flexible substrates and mechanically new options become available. Perhaps the most eye-catching commercial flexible devices today (2023) are curved and bendable displays which are - albeit still at a premium price - readily available [16, 17]. Less eye-catching, but nonetheless commercialized, are sensors based on organic materials. The company InnovationLab GmbH developed and is already selling organic sensor solutions[1].

Another well commercialized product are resettable fuses based on polymers. They heat up under power and limit currents. Current limitation is achieved by a strong temperature-resistance correlation of the devices. Such correlations can be used as a sensing principle. If the behavior of the device is known, a resistance can be translated to a temperature, rendering the device a

---

[1]`https://www.innovationlab.de/en/printed-electronics/products/`, accessed 19.06.23.

temperature sensor. As with most sensors, the measured variable (resistance) and the concluded parameter (temperature) are not the same. Furthermore, the first concluded parameter can be interpreted for further assumptions. In a heated device, the equilibrium temperature depends on the environment's ability to absorb heat. An environment with good heat absorption or conduction keeps the device cooler. The measured temperature can therefore be interpreted to obtain thermal information about the environment.

Elevated temperatures, however, are a major challenge for organic electronics. Further challenges are short lifetimes and quick degradation [18–20], both often linked with weak performance under elevated temperatures. Commercial materials are tested and optimized and can therefore be used right away in applications and devices (within their specifications). Lifetime and longevity are thus good. New lab materials, however, are often severely limited in their applications. Development of air-stable and long-living materials and devices remains challenging and the materials must be thoroughly tested. The world's largest chemical producer BASF, a partner of the 2-HORISONS project, created a promising semiconductor candidate for use in OFETs. The produced OFETs are intended for use in a sensor array. Sensors in an array are used to obtain information about an area rather than one single value. For technical reasons it is often useful to individually electronically switch through the single sensors of the array with transistors, making it an active matrix. The new semiconductor is tested for use in these transistors. As the semiconductor candidate is a new material, however, the fabrication of devices using this material requires tuning of process parameters. This can be done by simply guessing parameters and testing the respective devices, and then adjusting the parameters based on the results.

This work introduces a new sensor concept and application thereof and reports on a new organic semiconductor intended for use in OFETs. A brief explanation of the relevant topics as well as of the methods used is given in the beginning. Thereafter, a new sensor is introduced. First, the general sensing concept is described. The relation between resistance and temperature of the material is investigated and exploited for sensing. Temperature itself is not the relevant property but rather heat flow out of the device. To acquire data of an area, multiple single sensing pixels are then combined, and circuitry and a readout are added to the sensor to visualize the data. The finalized version of the sensor is then tested in two applications. In the first application, the sensor detects the height of a liquid inside a container. The information is obtained because of different cooling capabilities of materials and accordingly different temperatures of sensor pixels. Based on the required power to keep a specific temperature, liquid surfaces and interfaces are found. Due to gravity, the liquid's surface is flat and only the height is of interest. Therefore, the pixels are arranged in a vertical array. In the second application, the sensor is used to map airflow. It functions similar to the level sensor, but with varying airflow providing diverse cooling for pixels instead of the thermal mass of liquids. Airflow over a surface is usually not uniform. Because of that, the pixels of the sensor are arranged in a 2D array, resulting in the sensor mapping the airflow over a surface. Following the description of the sensor, the organic field effect transistor (OFET) is investigated. First, the material is examined with photoelectron spectroscopy (PES). The experimentally obtained chemical composition is compared with the desired structure to verify successful synthesis. Second, devices incorporating this material are investigated. Basic on/off functionality is tested first. The resulting on-off ratios as well as currents are reported to the fabrication team that then use these results to improve performance. After good samples have been produced, they are tested against operational stress. Devices are also continuously driven for days under various conditions, e.g. permanently on or permanently off. Finally, the temperature behavior of the devices is examined. As the OFETs are used in an active matrix in conjunction with temperature sensors, they must be stable over the whole temperature range to not falsify the results. Therefore, their behavior, like on- and off-current, is recorded for a temperature range of 20 °C to 90 °C. To estimate the influence of the OFETs on the result, they are also tested in series with the temperature sensors. The major findings are then summarized.

2. Fundamentals

2.1. Organic Electronics

Semiconductors are not restricted to metals and inorganic materials. Organic, Carbon-based materials can exhibit similar properties. They arise not from single atoms in a crystal but from molecules [21]. Physical/electrical properties of organic semiconductors are linked. A vast number of different molecules offer bespoke options for most needs [22].

To understand the electrical properties of organic materials, a C atom is considered. It has four valence electrons, two in the 2s and two in different 2p orbitals (Hund's rules). The two singly occupied orbitals can form bonds with other atoms with singly occupied orbitals. Energetically favorable, however, are four singly occupied orbitals, with one electron in the s orbital and three in p orbitals. Each of these electrons allows for a chemical bond. The s and p orbitals hybridize while forming bonds. Depending on the number of binding partners, different hybrids develop. In diamond, for example, each C atom forms bonds to four other C atoms. All four bonds are equal and in an angle of 109.5° [23]. They are sp^3 hybridized and very strong but also localized/fixed. Electrons can only move beyond these localized orbitals with very high energy, providing high electrical resistance [24]. However, if less than 4 atoms are bonded to the C atom, orbitals hybridize differently. In case of e.g. graphene, each C atom is bonded to only three atoms. The s orbital hybridizes with 2 p orbitals to three sp^2 orbitals, while the third p orbital remains. sp^2 hybrid orbitals overlap with (sp^2) orbitals of other (C) atoms and form a plane with an angle of 120° in between each orbital [24]. Perpendicular to the plane of these strong and localized σ orbital bonds is the remaining p_z orbital. p_z bonds of different C atoms overlap as well, forming electron occupied π and unoccupied π^* bonds. As this overlap is small, it is a weak bond and the electrons in it are not strongly localized. They form a delocalized π system over the molecule, given neighboring atoms with the same structure. This allows electrons to move within the molecule, breaking and reforming π bonds. Due to the hybridization and number of bonds and bonded atoms, single (σ) bonds and double (σ and π) bonds alternate. Such a system is called conjugation [25]. Hybridized molecule orbitals in the example of benzene are depicted in figure 2.1.

Hybridization lowers the energy of the molecular orbitals. Thus, the σ orbitals are lower in energy than the occupied π orbital. The latter is energetically the highest occupied molecular orbital ('HOMO') in ground state. The antibonding π^* orbital is above the HOMO and not occupied in ground state, forming the lowest unoccupied molecular orbital ('LUMO'). Being

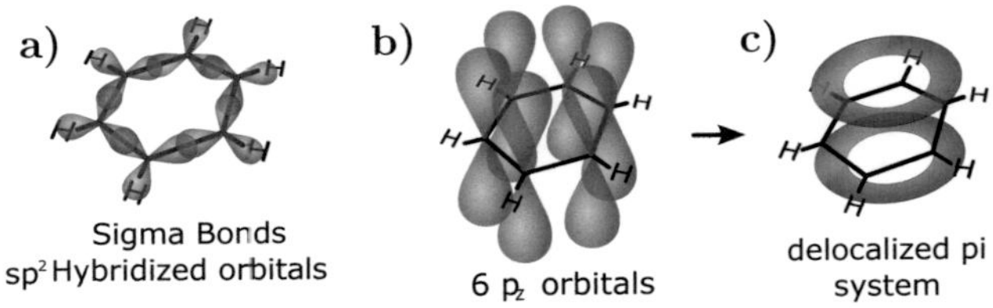

Figure 2.1.: Molecule orbitals in benzene. a) Planar oriented sp^2 orbitals overlap, forming strong bonds with localized electrons. b) The remaining p_z orbitals are perpendicular to the plane and weakly overlap. c) The weak overlap results in delocalized electrons over the entire molecule. Copied from [26] under CC BY-SA 3.0 and cropped.

influenced by their surrounding, the exact energetic position of the orbitals vary slightly in between molecules in a multi-molecule setting. Therefore, HOMO and LUMO don't have one exact value but rather a range. HOMO and LUMO are the organic analogue of valence and conduction band, respectively. Energetic distance between HOMO and LUMO is the band gap which is one of the most important properties of a semiconductor. It determines how much energy is required to lift an electron from HOMO to LUMO so it can potentially move along the molecule.

Although delocalized electrons in conjugated systems allow for charge transport in the molecule, charge carriers have to travel through many molecules. With few exceptions, organic semiconductors are dominated by high disorder [27, 28]. As such, electrons travel between molecules via hopping [28]: If wave functions of two molecules sufficiently overlap, there is a chance for an electron to 'hop' from one molecule to the other. This hop requires energy for the electron to be excited. The required energy is commonly derived from temperature, making hopping a thermally activated process [29]. Due to the necessary overlap of molecule orbitals, the spatial distribution and proximity of hopping sites to one another is of great importance for hopping probability. It has been shown that boundary conditions greatly affect charge carrier transport in organic semiconductors [30].

Similar to inorganic semiconductors, conductivity can be greatly enhanced by doping [31, 32]. Doping effectively increases the number of free charge carriers. Two contrary yet similar options are possible [33]. (Other doping mechanisms are possible yet not of relevance for this work.) Removing electrons from the semiconductor and binding them in a dopant creates an electron vacancy or hole. While the electron is fixed, the hole can travel through the conjugated system of the polymer. Alternatively, electrons can be added to the semiconductor. A dopant releases an electron into the polymer, creating an electron vacancy for itself. To successfully dope an organic semiconductor, the energy levels of semiconductor and dopant must align properly as sketched in figure 2.2. If p-type doping is desired, the dopant's LUMO must be at or below the semiconductor's HOMO. This alignment allows for the electron to transfer to the dopant due to an energetically

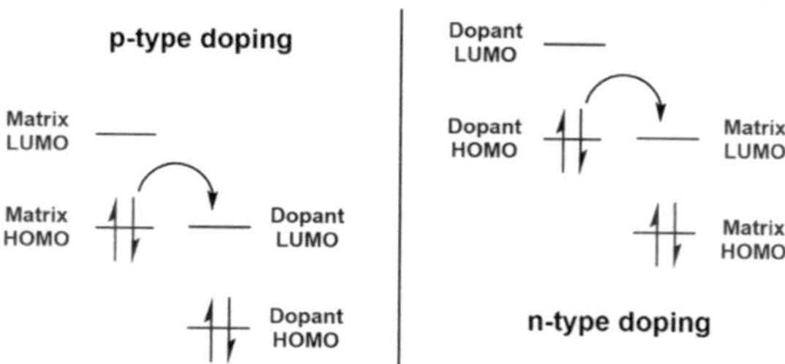

Figure 2.2.: Doping schematic for organic semiconductors. With the LUMO of the dopant at or below the HOMO of the polymeric matrix, an electron transfers from matrix to dopant and p-type dopes the semiconductor (left side). If the HOMO of the dopant is at or above the LUMO of the matrix, the latter gains an electron and becomes n-type doped (right side). Copied from [36] under CC BY 4.0.

favorable state. N-type doping requires the exact opposite. Oxygen in the air is capable of oxidizing polymers, causing unintentional p-doping and requiring countermeasures [34, 35]. Normally, however, doping is done in a well controlled and defined manner to increase conductivity of the semiconductor.

Proximity of molecules or charge conducting particles in general are of great importance in systems with only some conductive parts. One example of such a system are CPCs (conductive polymer composites). Conductivity in CPCs is reached by adding conductive particles to an insulating polymer matrix [37]. The conductive filler particles, which can be of organic or inorganic type, are blended with the polymer [38, 39]. Among others, size and concentration of the filler particles determine the spacing between them. In the case of insufficient conductive particles, they are isolated within the matrix and don't allow for charge transport between them. Increasing concentration of the filler decreases the gap between the particles, eventually bringing them in a close enough proximity that charge carriers can tunnel or hop (vide supra) between conductive particles. Thus, with increasing concentration of conductive fillers, conductivity of the CPC changes from insulating to conductive [37, 39–41]. The minimum concentration of filler to allow for conductivity is called percolation threshold. Figure 2.3 illustrates the relation between filler content and conductivity.

2.1.1. Organic Field Effect Transistor

The most important active electric component is the transistor [43]. 100 billion transistors in a single chip are common today, with large chips already exceeding the one trillion mark [44]. The simplest integration of a transistor as an on/off switch is simultaneously the most important, correlating to the digital bits 1 and 0. Outside of digital processing, the ability to switch and control electric voltage or current is also of great importance. It can, for example, turn the reading of a sensor on or off in an array of connected

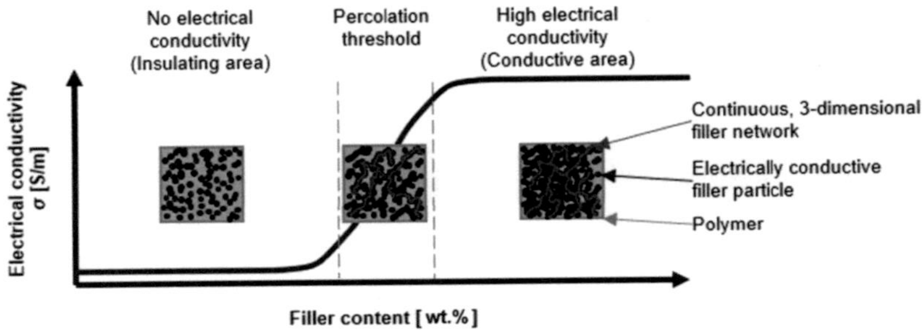

Figure 2.3.: Electrical conductivity of a CPC over filler content. The insets visualize the conductive filler (black spheres) and conductive paths in a polymer matrix (gray inset). Copied from [42] under CC BY 4.0.

sensors.

There are several types of transistors. The most used is the MOSFET type which stands for metal-oxide-semiconductor field effect transistor. While the term 'metal-oxide-semiconductor' describes the materials used, the term 'field effect' names the functional principle. The principle of field effect transistors has been successfully used with organic materials as well. Because of the organic materials, these transistors are named organic field effect transistor or OFET in short. For the electric component to be an OFET, only the semiconductor must be organic.

Figure 2.4 illustrates the schematic of a typical OFET. Source and drain are of the load circuit, the connection to be switched on and off. They are separated by a semiconductor. The control connection, called 'gate', is separated from the semiconductor by a dielectric. A voltage between gate and source can enable conductivity between source and drain. Most organic semiconductors are of p-type [45]. For that reason, a p-type semiconductor is assumed in the following description. The source is grounded and provides the charge carriers to travel through the semiconductor, hence the name. At the drain contact, charge carriers are drained out of the transistor when it is switched on. The gate opens and closes the channel through which charge carriers can traverse. Without an applied gate voltage V_G the organic semiconductor contains too few concentrated free charge carriers to allow for a (strong) current between source and drain. Applying a negative gate voltage $V_G < 0\,\mathrm{V}$ creates an electric field inside the semiconductor. Due to the electric field, holes are injected at the source and accumulate near the gate, driven by the field of the gate and stopped by the dielectric. They form a thin layer of free holes between source and drain. This layer is highly conductive due to the high concentration of holes and allows for a current to travel between source and drain upon applying a negative drain voltage $V_D < 0\,\mathrm{V}$. Switching off the gate voltage cancels the electric field and thus destroys the conductive channel. Remaining charge carriers and remnant conductivity can be reduced

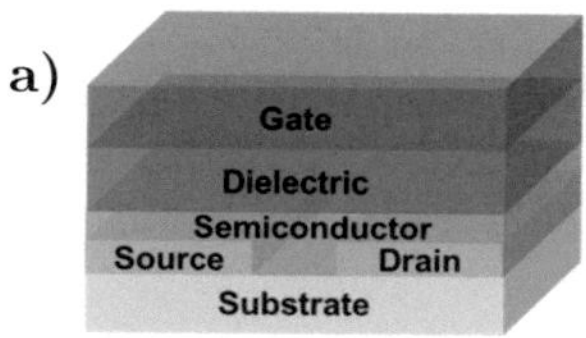

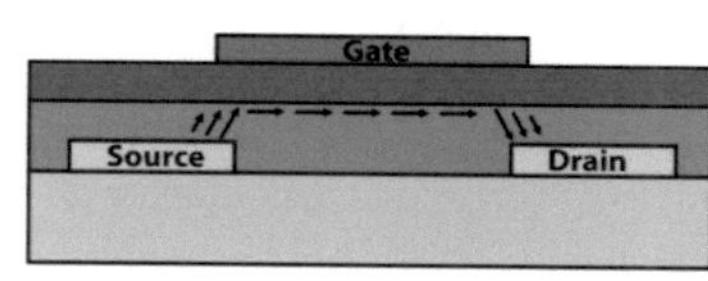

Figure 2.4.: a) Basic OFET structure. Copied from [46] under CC BY 4.0. b) Path of holes in a p-type OFET. Copied from [47] and adjusted under CC BY 4.0.

by applying a positive gate voltage. The electric field drives holes out of the semiconductor, creating a zone that is even further depleted. Thus, the gate opens and closes a conductive channel between source and drain. Because the gate is separated from the rest of the OFET by an insulator, there is no current through the gate. Therefore, an OFET is a switch with very low power consumption. Although source and drain have a different physical meaning in the scope of OFETs, they can structurally be similar, i.e. of the same material or dimensions. In symmetric devices they are interchangeable and definition is arbitrary or depends on the applied voltages. The principle can also be applied to n-type OFETs, but the sign of the voltages must be switched.

2.2. Positive Temperature Coefficient

Electrical conductivity of materials varies with temperature. If the resistivity increases with temperature, the material has a positive temperature coefficient, abbreviated as PTC. The opposite - lower resistance with higher temperature - is known as negative temperature coefficient or NTC. All metals are PTC [48]. This behavior has been utilized for temperature sensors. A well known example are platinum thermistors like the PT1000. They are precise and reproducible, standardized in several international standards, e.g. ASTM E1137 and IEC 60751. The PTC behavior is not limited to metals. Many CPCs exhibit PTC behavior as well [49–52]. They can be tuned/designed to exhibit a sharp PTC effect. This is achieved by mixing conductive fillers with a polymer with distinct temperature behavior. Distinct temperature behavior in this frame refers to morphological or phase changes, i.e. melting or change in crystallinity [53]. Below this distinct behavior temperature, the CPC is designed to be conductive. To achieve this, the filler concentration is just above the percolation threshold. As described above, this creates a conductive network of filler in the matrix. This network is susceptible, however. With increasing temperature, the structure of the polymeric matrix gets altered. For example, in the process of melting the composite, the previously transfixed position of molecules becomes mobile.The altered structure results in changes of the distribution of and distance between conductive filler particles. As

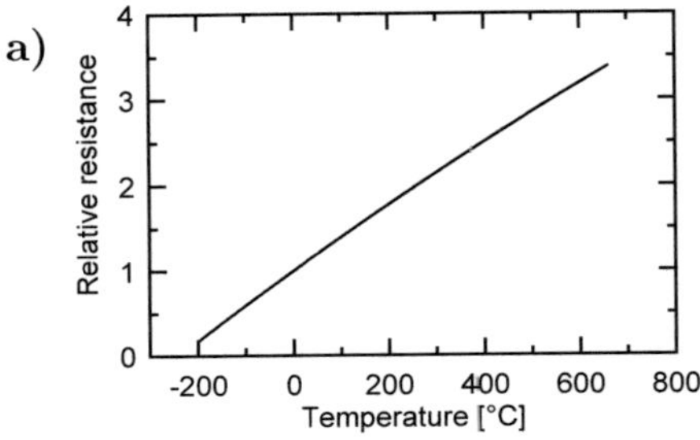 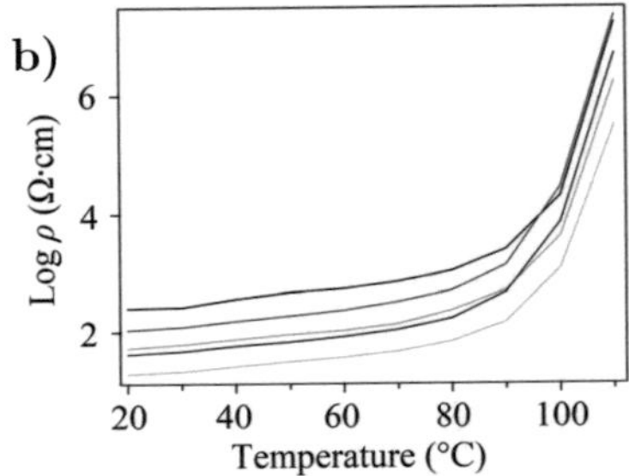

Figure 2.5.: PTC behavior in different materials. a) Resistance of a platinum resistance thermometer according to ITS-90 standard. The resistance is shown relative to the resistance at the triple point of water (0.01 °C). Data taken from [54]. b) Resistivity of a CPC system. The CPC shown consists of high-density polyethylene, carbon black and polyolefin elastomer. The latter is added for tailored mechanical properties. Copied from [55] under CC BY 4.0 and modified.

a result, conductive paths throughout the material system are destroyed and accordingly resistance increases. Contrary to a platinum resistance thermometer, the PTC behavior of CPCs is concentrated on the temperature of polymer change. This can be a broader temperature range, for example the softening of high density polyethylene, or a narrow temperature range, like in case of polymer melting. The narrower the distinct temperature behavior of the polymer, the steeper the PTC behavior of the CPC. In extreme cases this leads to an almost step-like PTC behavior where the resistance changes by a great factor over less than 5 K [53]. Exemplary resistance temperature correlations for metals and CPCs are depicted in figure 2.5.

2.2.1. LOCTITE ECI 8001 E&C

PTC behavior in CPCs has been known since the 1970s [56]. Since then, PTC behavior of CPCs has been artificially designed and tailored to suit different needs, with new CPC systems being researched to this day [57–59]. Research is mostly targeting resettable fuses, i.e. overcurrent protection [60]. They are also used as self-regulating heaters. For these applications, a sharp increase in resistance over a small temperature range is more desirable than a linear resistance-temperature correlation over large temperature regimes. The self-regulating effect arises from the power resistance dependence with constant voltage. An electric device with resistance R and voltage V draws power P according to

$$P = \frac{V^2}{R}.$$
(2.1)

In a resistor, all power P is converted to heat. Assuming constant voltage V, power P only changes with resistance R in inverse proportion. If the electric device consists solely of a material with PTC behavior, its resistance increases with temperature, and as a result, power draw decreases. Thus, a hotter device draws less power. This also reduces heating at higher temperatures, since the device heats itself solely through electric means. Given a pronounced PTC curve (and a reasonable voltage V), the device intrinsically cannot exceed a certain temperature.As this not only limits power draw of, but also current through the device, PTC materials are used as current limiters and resettable fuses [60, 61].

The commercially available heater inks from `Henkel AG & Co. KGaA` utilize this temperature induced power limitation. They are designed to exhibit a sharp increase in resistance over one order of magnitude in a small temperature span of only a few degrees and an otherwise relatively constant resistance. For `LOCITITE ECI 8001 E&C` - which was used in this work - the operation temperature is 65 °C [62]. At temperatures below the operation temperature, low resistance leads to high power consumption and thus strong electrical heating. Upon reaching the critical temperature and therefore strongly increased resistance, power consumption is significantly reduced, see equation 2.1. Since the change in resistance (manufacturer: PTC ratio > 7; the exact ratio will be determined in chapter 4.2.1) and the change in power are over one order of magnitude, heating is great below the operating temperature and low above. Thanks to this principle, the ink can be heated to its critical temperature and keeps the temperature stable over a range of voltages and nearly independent of its surrounding. As a commercially available material targeting self-regulating heaters, the manufacturer optimized reproducibility of the resistance versus temperature curve to ensure the same constant temperature independent of the history or number of previous uses of the device. Other advantages of the commercial ink are its ease of use and the well tested characteristics. `LOCITITE ECI 8001 E&C` is specifically designed for screen printing and comprised of a carbon filler and a polymer matrix, as well as additional components. The exact composition is the manufacturer's secret and not relevant for this work.

2.3. Heat Flow

In the previous section, a commercial ink designed as a self-regulating heater was introduced. The self-regulation only refers to active heating; it is not able to actively cool. That is not required for this work, however. According to the second law of thermodynamics, heat moves from a hot object to a cold object, if not otherwise forced. This effectively cools the hot object. Aside from niche mechanisms, heat transfer can be categorized as thermal conduction, thermal radiation, and thermal convection. In this section, conduction, radiation, and convection refer to their thermal parts unless otherwise noted. In the following, each mechanism is shortly introduced.

2.3.1. Thermal Conduction

Conduction is heat transfer within an object. According to the second law of thermodynamics, heat moves from hot to cold. Inside a solid object, heat is transported via free electrons and phonons. In metals, the main part of heat conduction originates from free electrons as there are plenty available. Due to their free nature, they can move fast and far through the object, carrying energy. Their movement is limited by scattering by other electrons, at crystal defects, or with phonons. The latter are collective vibrations of atoms or molecules in condensed matter. In solids with only few or no free electrons, phonons are the main contributor to thermal conductivity. The speed, quantity, scattering, and further properties of the phonons depend on the lattice of atoms and thus on the material. As a result, heat transport varies between materials. A material's ability to conduct heat is quantified by the thermal conductivity λ. (k and κ are also common symbols for thermal conductivity.) It is used to describe how much and how quickly heat is transferred due to a temperature gradient ∇T. The heat transfer is expressed as energy flow per unit area and unit time and named heat flux $\mathbf{q}$:

$$\mathbf{q} = -\lambda \nabla T . \tag{2.2}$$

The minus sign arises from the fact that heat moves from hot to cold, i.e. against the gradient. The equation describes the general case in three dimensions. For many applications it can be easier to consider the case of a long object with one hot and an opposite cold end and the other sides being thermally insulated. For example, air has a relative small λ and thermal contact with non-moving air can be considered almost insulated, validating the assumption above. To calculate the energy transfer or power through the object, the above equation can be simplified. Heat flux $\mathbf{q}$ is separated into energy per time $\dot{Q}$ or power $W = \dot{Q} = \frac{E}{t}$ per area A. ∇T simplifies to a temperature difference between the hot and cold ends over their distance d, i.e. $(T_{\text{hot}} - T_{\text{cold}})/d = \frac{\Delta T}{d}$. Together, this results in

$$\dot{Q}_{\text{cond}} = -\lambda \cdot A \cdot \frac{\Delta T}{d} \tag{2.3}$$

for conduction through an object with cross section A. Typical λ for metals - good heat conductors - are $237\,\text{W}/(\text{m·K})$ for aluminum, $401\,\text{W}/(\text{m·K})$ for copper and $429\,\text{W}/(\text{m·K})$ for silver [63]. Thermal insulators can have values as low as $0.0001\,\text{W}/(\text{m·K})$ for Mylar foil [63]. Common polymers are situated around $(0.1 - 0.7)\,\text{W}/(\text{m·K})$ (polyethylene terephthalate (PET) linear high-density: $\lambda = 0.52\,\text{W}/(\text{m·K})$ [64], polypropylene (PP): $\lambda = 0.22\,\text{W}/(\text{m·K})$ [65]).

2.3.2. Thermal Radiation

Contrary to conduction, radiation doesn't require matter to transfer heat. Thermal radiation transfers heat by means of electromagnetic waves. Due to thermal energy, particles inside matter oscillate. This causes dipole oscillations and accelerates charges which in turn create electromagnetic radiation. As everything above zero temperature has excited states and transitions, all matter above $0\,\mathrm{K}$ emits thermal radiation. Assuming a blackbody, i.e. a non-reflective and non-transparent body, in thermal equilibrium, the emission spectrum solely depends on the temperature of the blackbody and is described by Planck's law. The emissive power B is given by

$$B_\nu(\nu, T) = \frac{2h\nu^3}{c^2} \frac{1}{\exp\left(\frac{h\nu}{k_\mathrm{B}T}\right) - 1} \tag{2.4}$$

with frequency ν, temperature T, Planck constant h, speed of light c, and Boltzmann constant k_B. 'Black' in this context refers to the physical properties of absorbing all incident light. However, the body itself also emits radiation, appearing colored. Stars and the sun, for example, are often approximated as blackbodies. According to Planck's law (equation 2.4), the spectrum is temperature dependent. Total power can be obtained by integrating Planck's law over all frequencies. To further simplify, a flat blackbody surface emitting into a half-sphere is considered, resulting in the Stefan-Boltzmann law. Attributing for its surface A and imperfect blackbody emissivity ε, thermal radiation power of a surface is given by

$$P = \varepsilon \sigma A T^4 \tag{2.5}$$

with Stefan-Boltzmann constant $\sigma = 5.670 \cdot 10^{-8}\,\mathrm{W\,m^{-2}\,K^{-4}}$. ε describes the effectiveness of thermal radiation, ranging from 0 to 1. For $\varepsilon = 1$, the body is described as blackbody, while a body with $\varepsilon < 1$ is called gray body. $\varepsilon = 0$ describes a surface with no emission and is theoretically only possible for $T = 0\,\mathrm{K}$. For every other temperature T, the emission depends on the body's temperature.

Equation 2.5 describes the emitted power of a surface. The body, however, also absorbs thermal radiation emitted from its surrounding. Without a specific light source, the surrounding can also be assumed to be a blackbody, and thus emits radiative energy according to the Stefan-Boltzmann law. In total, the effective thermal radiation $\dot{Q}$ of a blackbody surface A with temperature T_body in an environment with temperature T_env is

$$\dot{Q}_\mathrm{rad} = \sigma \cdot A \cdot F \cdot \left(T_\mathrm{body}^4 - T_\mathrm{env}^4\right). \tag{2.6}$$

View factor F describes the different surface areas of the environment and the body. Since the effective radiation (as energy loss of the blackbody) accounts for absorption of the surrounding's emission, the temperature of the environment is important for the quantification of $\dot{Q}_\mathrm{rad}$.

2.3.3. Thermal Convection

Convection is heat transfer by movement of a liquid or gas. Heat is transported by being absorbed by a medium and subsequently moved with the medium.

2. Fundamentals

Most coolers rely on thermal convection as part of the process. Common examples include air or water coolers. In these cases, the medium that transports the heat is subjected to forced movement. It is thus called forced convection. Forced movement, however, is not required for convection as convection also happens naturally. Driving forces, in that case, are of passive nature, and density and pressure related. As the convection medium, for example air, in contact with a hot object absorbs thermal energy, its density decreases. This causes the hot air to be displaced by colder, denser air, and moved upwards. The cold air in contact with the hot object is now heated, while the hot air can dissipate its heat to its surrounding and cools down. Therefore, the medium is in motion without being forced by a motor or pump. This unforced convection is called natural or free convection. Mathematical descriptions of all the physics behind convection, both natural and forced, are complex and require profound knowledge of properties of, among others, convection medium, body, and geometry [66–68]. For the sake of simplicity, Newton's law of cooling is used [69]. Heat transfer by thermal convection through surface A is thus described as

$$\dot{Q}_{\text{conv}} = h \cdot A \cdot \Delta T \qquad (2.7)$$

with temperature difference ΔT between object and environment. The equation uses heat transfer coefficient h, which summarizes all properties of the system without knowing individual values. For common situations and mediums, an estimate of h can be looked up for similar situations [70]. Free air convection has an h of $(2.5 - 25)\,\text{W}/(\text{m}^2\,\text{K})$, forced convection is in the range of $(10\text{-}500)\,\text{W}/(\text{m}^2\,\text{K})$.

2.3.4. Total Heat Flow

In real world applications, the presented mechanisms seldom act alone. All mechanisms work simultaneously, to different degrees, governed by equations 2.3, 2.6, and 2.7. It is therefore reasonable to summarize these equations to the total heat transport equation

$$\dot{Q} = A \cdot \left(\varepsilon \cdot \sigma \cdot \left(T_{\text{body}}^4 - T_{\text{env}}^4 \right) + \left(T_{\text{body}} - T_{\text{env}} \right) \cdot \left(h + \frac{\lambda}{d} \right) \right) . \qquad (2.8)$$

Equation 2.8 describes the total effective heat flux out of a body with temperature T_{body} in an environment with temperature T_{env}. Surface area A and the temperatures are relevant for all mechanisms. Apart from that, the heat transport mechanisms function independently. Thus, as long as temperatures remain the same, variations in one heat transport mechanism do not influence the others ones.

3. Device Fabrication and Characterization

In this chapter the fabrication method of devices designed and used in this work is described, and methods and setups of experiments are introduced

3.1. Screen Printing

The sensors used in this work are screen printed. Screen printing is a fabrication technique where a coating is applied to a substrate through a screen. Figure 3.1 illustrates the process. The eponymous screen consists of a fine mesh tightened in a frame. The mesh is coated with a (UV) light sensitive coating. Light is then used to cure/harden the coating. Between light and screen is a mask with the negative of the desired pattern. Thus, not all of the coating is cured, and the uncured coating is washed off the screen. The result is a screen with the desired pattern. This screen is placed over a substrate. On one end of the screen, ink is applied in a line. The ink is then pushed over the screen with a squeegee. In places where the mesh is not covered by the coating, ink can traverse the screen and reach the substrate. In places where the mesh is covered by the coating, no ink can traverse the screen and the substrate remains blank. In this manner, the ink forms the negative of the impermeable coating pattern on the screen. The whole process is time efficient, as it only requires the squeegee to be moved over the screen to produce an ink pattern. Afterwards, the substrate can be removed, the ink is usually pushed back on the frame to the beginning line, and the process can immediate be repeated with the next substrate, allowing for good throughput. Because the ink is only applied to the substrate where needed and otherwise used for the next substrate, ink losses are minimal. Although the lifetime of a screen is limited, it can be used for many runs and a long time (depending on the exact specification). In summary, screen printing offers an efficient and cheap method for medium to high quantities of printed electronics.

In this work, the devices in chapters 5 and 6 were fabricated by screen printing. The screen printer used is a `Thieme 3010S` located in the clean room of the facilities of InnovationLab. Polyethylene naphtalate (PEN) (Teonex Q51) was used as substrate. It was cut to size and heated to 120 °C for 1 h prior to printing. Polyimide (PI, 'Kapton') substrates were also used to test mechanical flexibility. For printing, ink was applied directly on the screen. The screen then was first flooded with 300 mm/s and immediately afterwards printed with 400 mm/s. Squeegee pressure and angle were set to 2.3 bar and 73 °C, respectively. The first few (≈ 2) runs with each ink were discarded as the films had holes and defects in them. Thereafter, ink was properly distributed over the screen and the printed pattern complete. For

3. Device Fabrication and Characterization

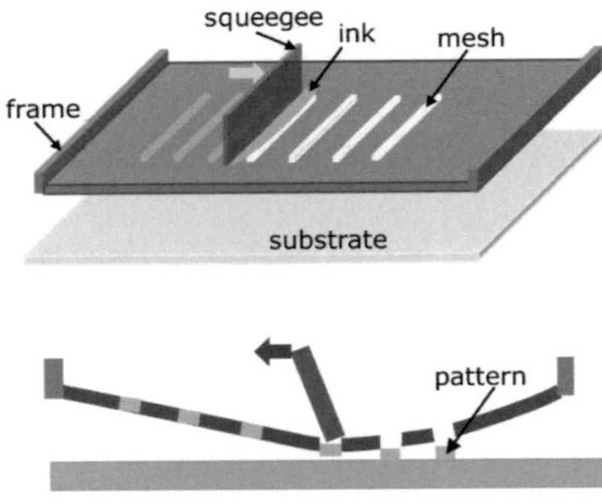

Figure 3.1.: Schematic of screen printing. Copied from [71] under CC BY 4.0.

this evaluation, optical investigation with a magnifying glass was sufficient. After each printing step followed an annealing step, partially split into drying after print and main annealing en bloc for all substrates simultaneously. For alignment the first layer has position markers outside the sensor matrices that can be detected by the camera system of the Thieme printer. These markers together with markers on the screen itself allow for precise alignment of all layers. The screen printed devices consist of up to six layers. These, from top to bottom, are silver for circuit paths, dielectric for insulating bridges, another layer of silver for bridges over the dielectric, a PTC ink, a low resistive ink and finally a high resistive ink. The last layer is not present in the level sensor.

Screens were ordered from Hans Frintrup GmbH. They have a total size of 677 mm x 690 mm and use a webbing of PET with a mesh specification of 130-34 Y. The emulsion over mesh (EOM) is 6-14 µm. For circuit paths silver ink LOCTITE ECI 1010 E&C from Henkel AG & Co. KGaA was used. It is designed as conductive ink for screen printing of flexible circuits [72]. This ink was dried for 2 minutes at 80 °C immediately after printing. Once silver ink has been printed on all substrates they were all annealed for 10 minutes at 80 °C. This procedure deviates from the recommended one in the technical data sheet (TDS) [72] but had been employed successfully for some time by the InnovationLab GmbH. To allow for crossings of conducting lines, the dielectric ink LOCTITE EDAG PF 455B E&C was printed over the bottom line [73]. Following recommendations, this layer was printed twice with it being UV cured in between and after. Two layers are required to ensure full electric separation between conducting lines above and under. On top of the dielectric, another layer of LOCTITE ECI 1010 E&C was printed to connect the crossing electric line. LOCTITE ECI 8001 E&C is designed for self regulating heaters [62]. It features a PTC behavior based on carbon fillers with an operating temperature of 65 °C and a PTC ratio of >7. Curing followed the procedure of the silver ink, albeit with increased temperatures of 90 °C both for drying and annealing. The low resistance layer consists of pure LOCTITE ECI 7004 E&C [74]. This carbon based ink is intended for use in force sensitive modules or temperature sensors. Though this is not the intended use in this work it was utilized because of the experience with this material at the facilities of InnovationLab[1]. This allows for educated guesses of the resulting resistance already in the design stage of the devices without extensive characterization beforehand. Drying and Annealing for this ink was the same as for the silver ink. Contrary to the other inks the high resistive layer is a mixture of two inks, namely 40% LOCTITE ECI 7004HR E&C (low resistive layer, vide supra) and 60% LOCTITE EDAG PM 404 E&C. The latter is non-conductive and intended to alter the resistivity of its corresponding conductive ink [75]. By mixing it with the conductive ink used for low resistive layers the mixture yields a high resistance. For the use-case in this work the exact resistance is of little importance as long as it is high and consistent over all substrates and pixels in one device. This ink was printed and post treated the same as the silver ink. All inks were thoroughly stirred for several minutes prior to use but otherwise used as received.

To ensure good thermal contact between the sensor devices and the surface underneath thermal paste was applied in some instances. The thermal paste in this work is RS-GCS-040-GNS silicon free grease from RS Components. As it is non-conducting a small amount was applied directly onto the PTC element of the circuit with a toothpick.

3.2. PTC Sensor Array Readout

Readings of the full sensor devices are taken with an Arduino-compatible Mega 2560 R3 board. Arduino is an open-source multi-purpose microcontroller. Importantly for this work the Mega variant features an onboard 16 channel ADC (analog-to-digital converter). The 10-bit ADC works for a range between 0 V and 1.1 V, 2.56 V or 5 V. (An upper limit (<5 V) can also be set by feeding the voltage to the board; due to the required additional voltage supply or circuitry this was not used.) Upper limit is set manually during operation depending on the read values. The ADC was used to read the voltage over the shunt resistor (reasons given in chapter 4) of the individual pixels of the sensor arrays in chapters 5 and 6. Each line of a pixel was connected to an analogue input. Grounds of readout and device were connected as well. Values were read one by one and then sent together serially via USB connection to a computer which displayed and stored the data. Not optimized for speed, this configuration as used handled a reading speed of around 50 Hz. The self-written codes used are shown in the appendix A.1.

---

[1]Personal communication with Dr. J. Zimmermann and B. Obermayer, both InnovationLab GmbH.

3.3. Film Preparation

Films for XPS were prepared on a Si substrate. Substrates were cut from a wafer of polished intrinsic silicon with a native oxide layer to ca. $13 \times 13\,\mathrm{mm}^2$. Afterwards they were cleaned in an ultrasonic bath with 10 min each acetone and isopropyl alcohol (IPA) and dried with a nitrogen gun. The XPRD01B06 solution was received from S. Raths (BASF) and stirred for at least 10 min at 60 °C before use. It was then spin coated onto the substrates at 1000 rpm for 60 s. Directly afterwards, the samples were annealed on a hot plate at 110 °C for 1 min. Both spin coating and annealing were carried out in a glovebox with nitrogen atmosphere. The samples, however, were in ambient conditions before measurement.

3.4. Analytical Methods

3.4.1. I-V Curves and CV Measurements

I-V curves and CV (constant voltage) measurements were recorded with an `Agilent 4155C`. It features 4 source measure units (SMUs). These were used for both PTC based sensors and OFETs. One SMU was set either to a voltage sweep, e.g. producing a linear voltage curve from $+10\,\mathrm{V}$ to $-30\,\mathrm{V}$, or to a constant voltage for a set time, as appropriate. The other SMUs were set to constant voltages or zero currents and recorded current and voltage. SMUs were connected via triax cables to a probe station and needles. Inside the probe station is a hot plate controlled by an `Intertek 1000` temperature controller from `Intertek ASG`. The hotplate was electronically heated up to 100 °C, cooling was passive. This allowed the investigation of temperature behavior of OFETs and PTC samples.

3.4.2. Photoelectron Spectroscopy

Chemical analysis was carried out via photoelectron spectroscopy (PES). Due to its complexity, method and underlying principles are briefly summarized in the following. For further details, the reader is referred to the *Handbook of X-ray Photoelectron Spectroscopy* [76] which this explanation largely follows.

Both name and method of photoelectron spectroscopy are based on the photoelectric effect. Atoms subjected to electromagnetic radiation can absorb the energy. This lifts an electron to a higher energy level. If the absorbed energy is higher than the energy keeping the electron near the atomic nucleus, it is ejected off the material [77]. Figure 3.2 illustrates the photoelectric effect. The minimum energy required for this process is called ionization potential and is the sum of the binding energy $_{\mathrm{bin}}$ and the work function ϕ. Binding energy E_{bin} is given relative to the Fermi level E_{F}, the difference between vacuum level E_{vac} and Fermi level E_{F} is the work function ϕ. (As a note: in PES energy levels are usually referenced to $E_{\mathrm{F}} = 0$. For that reason, the

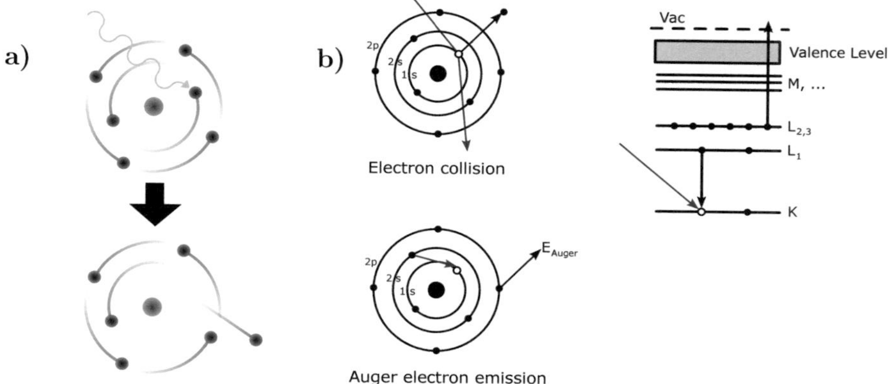

Figure 3.2.: a) Photoelectric effect. An incident photon (sinuous line) hits an electron. Center sphere is the atomic core. The electron absorbs the energy of the photon and is emitted. Copied from [78] under CC BY-SA 3.0. b) Auger effect. Left side shows an inelastic electron scattering ('Electron collision', top) exciting a 1s electron. A 2s electron falls in the 1s vacancy and the created energy emits a 2p electron ('Auger electron emission', bottom). Right side shows the same process in energy levels. Copied license free from [79].

ionization potential is the sum of the binding energy and the work function. Outside from PES binding energy and ionization energy are mostly synonym.) Therefore, the incident photon's energy $h\nu$, with Planck's constant h and the photon's frequency ν, has to be higher than the sum of the electrons binding energy E_{bin} and work function ϕ combined:

$$h\nu = E_{\mathrm{bin}} + \phi + E_{\mathrm{kin}} . \tag{3.1}$$

Any additional energy is converted to kinetic energy E_{kin}. Therefore, with known incident energy and work function, measuring the kinetic energy yields the binding energy. Electrons below the Fermi level cannot escape the material. In bulk material with many atoms, electrons separated from their nucleus can interact with other electrons by collisions and scattering. Because of inelastic scattering and resulting energy loss, separated electrons have energies between their initial and zero kinetic energy. At zero kinetic energy, electrons are not emitted off the surface. The instrument detects electrons down to this energy, upon which detection cuts of. Consequently, it is called the secondary electron cutoff (SEC).

Emitting an electron leads to a vacancy or 'hole' in the atom. This gets filled by another electron at initially higher level, releasing energy. The energy can be transferred to another electron of the same atom, potentially emitting it. Due to the discrete energy levels in an atom, electrons emitted this way have energies specific to the element. This process is called Auger effect or Auger-Meitner effect after its discoverers [80, 81]. It is illustrated in figure

3.2.

The incident energy is determined by the source of radiation. For this thesis a Helium gas discharge lamp with ultraviolet light (ultraviolet photoelectron spectroscopy, UPS) and X-rays (X-ray photoelectron spectroscopy, XPS) from aluminum anodes are used. The Helium gas discharge lamp features two main emissive energies, one at 21.22 eV, denoted He I, and one at 40.60 eV, denoted He II [82]. Only the first, He I with 21.22 eV, was used. Due to the high intensity of this radiation, UPS features high sensitivity with low signal-to-noise ratio. It is especially suitable for determining the work function ϕ via the SEC. The precise position of the SEC is calculated by extrapolation its linear regime to the flat background. For technical reasons, namely a potential between sample and analyzer, an acceleration voltage of roughly 5 V is applied. The precise value will be determined with a reference sample with known energy levels and is not of relevance. XPS on the other hand uses much higher energies than UPS albeit with lower intensity. For X-rays from aluminum anodes as used in this work the incident photon energy $h\nu$ is 1486.6 eV. This allows to transfer more energy to electrons so deeper bound states, i.e. core-level states, are accessible. The core-levels are characteristic for the element and sensitive to charge around the atom and its bonds. The *Handbook of X-ray Photoelectron Spectroscopy* provides an overview over core levels for each element and potential chemical shifts of various bondings. This allows not only for identification of elements but also for estimation of chemical bonds.

PES is not only suitable to identify atoms within a surface but also to quantify atomic ratios. The peaks in the spectrum can be accounted to the element. Electrons that scatter only elastic or not at all have the full kinetic energy as described by equation 3.1. A macroscopic sample with many atoms yields many electrons with their respective full kinetic energy according to their binding energy. Some of those electrons scatter inelastic and have reduced kinetic energy. This results in a peak in the energy spectrum with a trail to lower kinetic energies. Its intensity depends on various things, many of which are instrument specific. Others factors are the number n of atoms of the specific element and the photo ionization cross section σ. The detected intensity I is

$$I = n f \sigma \theta y \lambda A T \,. \tag{3.2}$$

λ is the mean free path and y is the formation efficiency of photoelectrons. f (x-ray flux), θ (angular efficiency factor), A (area of observed interaction), and T (detection efficiency) are all instrument dependent. More in depth explanations to this equation can be found in [76, 83].

Most of the parameters are intrinsically constant or can be made independent of the sample by the right setup. The varying parameters are only n, I, and σ. Thus, the number of atoms can be expressed as

$$n = I / \left(\sigma \cdot c \right) \tag{3.3}$$

with the constant c summarizing all the equipment parts of equation 3.2. While the absolute number of atoms cannot be obtained without knowledge of c, ratios of atoms can be calculated from intensities with

$$\frac{n_1}{n_2} = \frac{I_1/\sigma_1}{I_2/\sigma_2} \, . \tag{3.4}$$

The photo ionization cross sections of atom orbitals for different excitation energies is available in literature [84]. Therefore, atomic ratios can be calculated from peak intensities and the corresponding cross sections.

The spectra of this work were recorded with a PHI Versa Probe II system with a concentric hemispherical analyzer as described in [83, 85, 86]. Sources were an Omicron HIS 13 helium discharge lamp for UPS and a monochromatic Al Kα anode for UPS. X-ray source and analyzer have a 'magic angle' to one another to allow for simplifications of equation 3.2. Prior to each measurement, a freshly sputter cleaned Ag foil was recorded to determine the Fermi level. Samples were prepared in a glovebox, vide supra, and carried through normal (clean room) conditions to the equipment.

3.4.3. Dektak

Layer thicknesses were recorded with a stylus profilometer. For this technique a diamond stylus moves over the surface of a sample. The profile of the sample determines the stylus' vertical displacement which is translated by the equipment into digital data. Recordings were taken over a straight line across the sample with appropriate length. Samples were tested multiple times and averaged. The equipment used is a Dektak 150 from Veeco Instruments, Inc..

3.5. OFET Fabrication

OFETs for characterization in chapter 7 were fabricated by employees of InnovationLab GmbH (B. Obermayer, S. Schlisske, M. Seifert, K.-P. Strunk) and BASF (S. Raths). Contact structures were fabricated via lithography and the active layers were screen-printed with the Thieme 3010S mentioned earlier. Only complete OFETs, as single test structures or fully integrated within a matrix, were provided for characterization. The author of this work was not actively involved in the fabrication. The fabrication process was adjusted several times after feedback from the characterization, upon which new OFETs were produced and analyzed.

A rough design schematic of an OFET with downstream sensor in an active matrix is illustrated in figure 3.3. The dimensions and number of fingers within the finger structures were varied for characterization and optimization. All contacts are made of gold because of its energy levels. Source and drain are on the bottom, the gate is on top (top-gate bottom-contact). Onto the OFET contact structure, the new semiconductor XPRD01B06 was printed

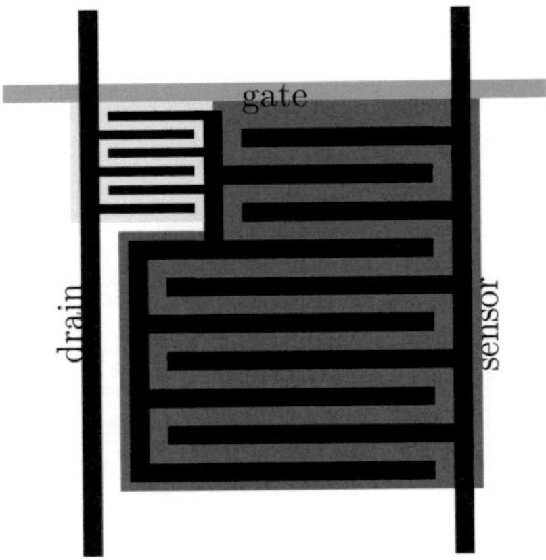

Figure 3.3.: Schematic of one pixel within the active matrix. Very light gray area in the top left is the OFET. It is also the area covered by the gate electrode (not shown). The area of the sensor is depicted in dark gray. Contact lines are labeled with their connection.

and afterwards encapsulated by a dielectric. The gate was then printed on top of the dielectric. As final step the sensing material was printed onto the sensor structure. The source is directly fed into the sensor.

For OFET characterization, isolated OFETs with contact lines were prepared. If necessary, the substrate was cut to size and inserted into a chip holder for easier connection to the setup. In the active matrix, drains from all OFET are connected together, gates are connected across, and sources are connected down. To differentiate between OFET and sensor effect, some test matrices have a connection to the source, in between OFET and sensor. Due to additional contact lines and the required space, these matrices are limited to 4x4. The biggest matrices tested in this work have 32x32 pixels.

4. Sensor Concept, Layout, and Verification

Technologies of the last decades and centuries have aimed to help people in their everyday life, as well as in highly specialized scenarios. With the rise of computing power and even more with artificial intelligence (AI) many things are controlled by some kind of electronics. They work by observing the current status and acting accordingly. To properly function there is high need for various sensors and actuators. In the fast paced world of today it is highly beneficial for these sensors to be simultaneously general in concept and adaptable and customizable in design and exact specifications. Scientists and researches all over the globe are expanding the portfolio while system integrators and manufacturer are always looking for new sensors [87–89].

While there are many types of sensors, almost exclusively the device registers a property different of the desired information and interprets the result to return information about its environment or applied device. Examples are a common Pt100 thermistor to measure temperature, while in fact its resistance is measured and converted to temperature, e.g. according to EN 60751. Another example is a digital ammeter, as it measures the voltage drop over a precise shunt resistance and calculates current. Similarly, in this work, physical properties of the sensor are utilized to draw conclusions about different system properties. Energy flow from an electrically heated device to its surrounding and a change in resistance as a result is the identifier and will be interpreted to sense several attributes of its environment.

To satisfy cheap fabrication and effortless customization the sensor is fully screen printed. Screen printing 'has been proclaimed to be the best among the other additive manufacturing techniques' according to Suresh *et al.*[90]. Details about screen printing and sensor fabrication are given in chapter 3. As commercial inks with the desired properties are available there was no need to design and verify a unique ink for the purpose of the sensor developed in this work. Therefore, the individual components of the device are already well established and the reliability and reproducibility are inherited.

This chapter aims to describe the basic concept and verify its working principle. In the chapters thereafter two different applications are demonstrated, namely a liquid level sensor (chapter 5) and a surface airflow sensor (chapter 6). The explanation in this chapter as well as the application in chapter 5 follow in large parts the procedure described in the paper about the level sensor [91].

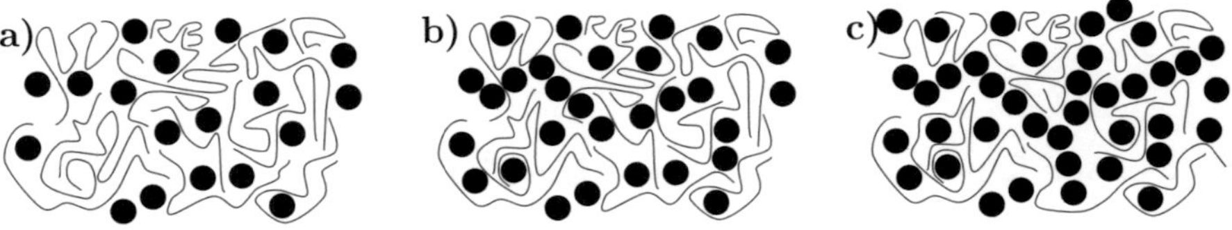

Figure 4.1.: Illustration of impact of filler concentration on conductivity of CPCs. Lines represent the polymer, black spheres represent conductive filler. Concentration of the filler increases from low in a) to medium in b) and high in c). Compare figure 2.3.

4.1. Conception

4.1.1. Sensing Concept

Most important component of the sensor is a CPC. CPCs consist at least of a highly conductive filler and an insulating polymer matrix [41, 92]. While the earliest CPCs incorporated only two components, newer works can contain many different components [93]. Each component contributes to the attributes of the composite in a different matter, influencing i.e. mechanical and/or thermal properties [94]. Conductivity of the CPC is determined by a conductive filler and its density. Examples of filler are carbon black, carbon nanotubes, and metal microspheres. At low filler concentrations, agglomerates and particles are isolated within the matrix. As they lack connection in between, charge carriers cannot be transported over large range. They may create hot spots of local electric conductivity but no long ranging networks. Thus, their contribution to conductivity is low and resistivity is governed solely by the polymer. Most polymers are insulating in nature, and so are CPCs with low filler concentration. However, by increasing the concentration the spacing between filler particles decreases. This is illustrated in figure 4.1 and in the insets of figure 2.3. Filler particles in close contact allow electrons to tunnel from one to another and form conductive paths through the filler. At a specific concentration there is just enough filler material inside the polymer for conductive paths to range through the whole bulk uninterrupted. The minimum concentration needed to form this large ranging network is called percolation threshold [92]. Thus, conductivity arises from paths of conductive filler particles distributed throughout the entire CPC. As there is already a network of conductive paths throughout the material at or above the percolation threshold, additional filler content has diminishing effect on the conductivity which is visualized in figure 2.3. Charges can travel through existing conductive paths and don't benefit from further percolation options. The percolation threshold is usually sharply defined over concentration [95–99]. One example from literature is depicted in figure 4.2.

Because of the strong change in resistance with changing concentration, the conductivity of the material is highly sensitive to filler concentration near the percolation threshold. A blend of materials with different coefficients of

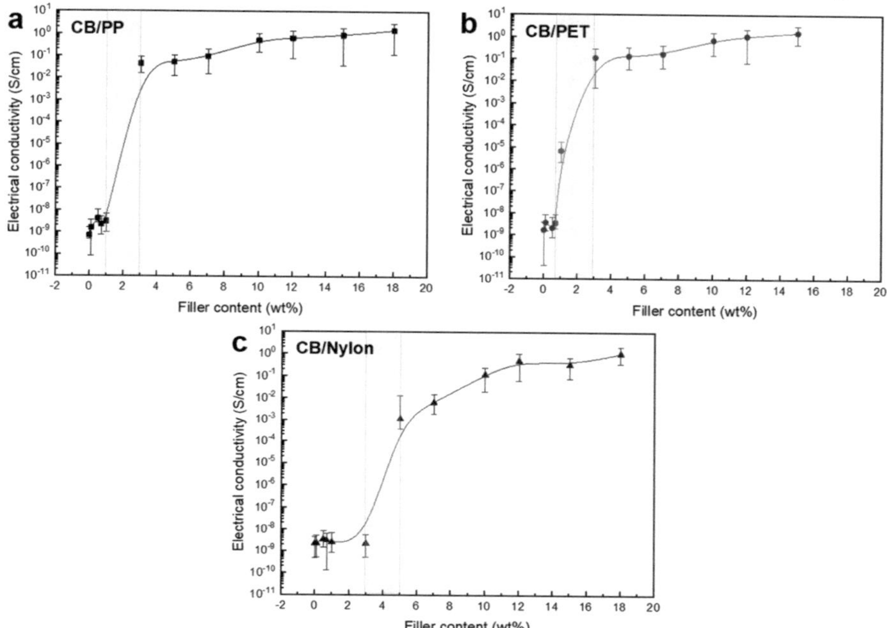

Figure 4.2.: Exemplary percolation threshold in literature. Taken from [97] under CC BY 4.0.

thermal expansion α changes the volumetric concentrations of its components. Polymers typically have a higher α. Therefore, heating a CPC decreases its volumetric filler concentration, resulting in lower conductivity. As such, CPCs are PTC materials. Other internal changes, i.e. of morphological nature, can also influence filler concentration, also locally. The high sensitivity in conjunction with volumetric concentration are the key components of the sensor concept in this work.

The PTC behavior of CPCs is typically step-like [50, 52, 53, 96, 100–104]. Starting point of the following explanation is a CPC with a filler concentration just barely above the percolation threshold. Due to the conductive network - established with sufficient filler material - the CPC is conductive. Depending on the coefficient of thermal expansion α, increasing temperature and resulting expansion can disrupt some conductive pathways leading to PTC behavior. Heating to the temperature of a phase transition, for example melting point or glass transition, the molecular structure of the polymer changes. This change in the arrangement of the polymeric matrix disrupts the conductive network, increasing resistivity [105]. As the CPC is at the percolation threshold a small change in molecular structure and thus filler spacing and/or distribution has a huge impact on conductivity. With the phase transition usually occurring over a small temperature window, resistivity changes step-like with temperature in this regime [51, 53]. This grants the resistance of the material a very

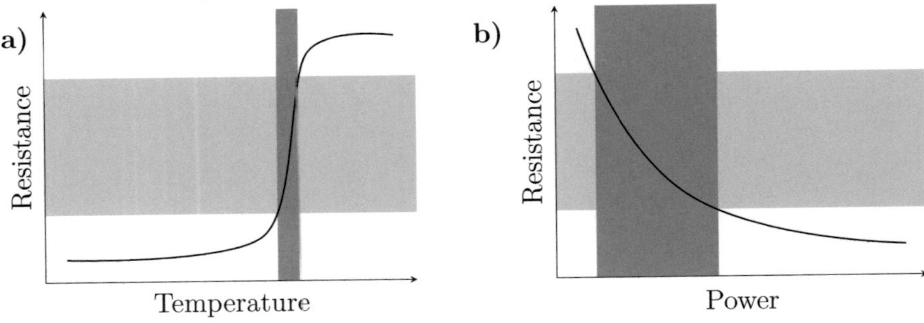

Figure 4.3.: Self-limiting heater effect of PTC material. a) Resistance exhibits a step-like behavior over temperature. b) Resistance and power are coupled by $P = V^2/R$. A small temperature range (dark gray in a) results in a large resistance change (light gray in a and b), which in turn results in a large change in power (dark gray in b).

high temperature sensitivity around the temperature of the phase transition. With temperatures above the phase transition the conductive network is destroyed and the conductive filler contribute only little to conductivity. It correlates to a CPC with low filler concentration, vide supra. Resistivity is accordingly high. It is mostly governed by the polymer and thus nearly temperature independent again. In summary, the resistivity of CPCs is low at low temperatures and high at high temperatures, with a step-like change over a small temperature range in between.

The resistance-temperature dependence allows for self-limiting electric heaters [51, 106]. Power consumption in electrical devices follows equation 2.1. In the PTC CPC, all of the electrical energy is converted to heat. Thus, with an applied voltage V_in, the PTC material heats up. Providing enough electrical energy, the temperature of the CPC reaches the phase transition temperature of the polymer of the CPC. At the phase transition, resistance R_PTC of the PTC material drastically increases as described above. According to equation 2.1, P is inversely proportional to R. Heating power P decreases accordingly. Following the step-like PTC behavior, power draw decreases also step-like, effectively switching off heating. Therefore, the CPC is heated below and not heated above the temperature of the phase transition, effectively staying at constant temperature. This temperature is labeled operation temperature T_op from here on. Relations of the properties are sketched in figure 4.3. Power P also depends on the voltage V. Eliminating this variable by setting it constant makes the heating power vary only with R. The voltage can be chosen such that the electrical power is sufficient to increase material temperature at low resistance, i.e. below T_op, but not enough to increase it at high temperature, i.e. above T_op. Generally all PTC materials can function as self-limiting heater. However, only the step-like characteristic of CPCs allow for a precise operation temperature T_op defined by material properties. It reaches and then safely maintains T_op.

For the self-limiting heater to function as such T_{op} must be above ambient temperature T_{am}. This is achieved by choice of material and setup of experiment: during operation the heated PTC is at higher temperature than its surrounding. With positive temperature difference, heat is transferred from the heated PTC, see equation 2.8. Upon reaching T_{op}, temperature ceases to change and steady state sets in. As temperatures stays constant, outgoing heat flow and electric heat generation are of equal quantity. The electric heat generation can be quantified by measurement of the electric current I and calculating the power by $P = V * I$. Consequently, the electric power serves as a measure for heat transfer from the PTC to its surrounding. Assuming constant temperatures T_{op} and T_{am}, the outgoing heat flow can be used to make assumptions about the PTC's thermal environment as described in chapter 2 in details. T_{op} is constant due to the material, T_{am} can be assumed constant if heat flow is small and thermal mass of the surrounding is relatively high. The parameters in equation 2.8 are either defined by the test system (for example A as the area of the active PTC element) - and thus constant - or the immediate surrounding of the heater (h and λ). As h and λ are properties specific to certain surroundings (e.g. λ is a property of the material in contact with the PTC), the PTC's power draw depends on its environment. Therefore, various (states of) surroundings result in different heat flux and the PTC can be used as a sensor to differentiate between them. To verify the suitability of the principle, two applications of this sensor are given in chapters 5 and 6 where it is applied as a liquid level sensor differentiating materials and as a surface airflow sensor where it detects air speed, respectively.

The principle is ideal to detect differences and not absolute values. Absolute values require more precise data and further processing, including emissivity σ, ambient temperature T_{am}, and proper 3D integration, and they are also not the aim of this work. Rather, the focus lies in differentiating between surroundings with simple design. A single PTC element can already signal dynamic change. However, with one reading alone little information about the surrounding at only one spot is gathered. Instead, areal resolution is desired, inspired by the work of the 2-HORISONS project, which includes, amongst others, active and passive sensor matrices. For that, several PTC elements are to be applied simultaneously, with each yielding information about a different location. However, connecting several PTC elements and reading their resistance individually poses some challenges. These are resolved by the circuitry described in the following.

4.1.2. Circuit Design

The operation temperature of the PTC is not reached instantly when voltage is supplied. It instead takes a couple of seconds. The exact times will be investigated during tests, vide infra. Because of the heating period it is not feasible to heat one element to its operation temperature T_{op}, read the electric power, and proceed with the next one. This would have devastating impact on response speed, especially if many PTCs are to be deployed. Consequently, heating and reading of individual elements is separated. For this a shunt resistor R_{SR} is connected in series with each PTC element, forming a voltage divider. In a voltage divider, the relative resistances of both components can be determined from the voltage over the whole voltage divider and the voltage in between the two resistors. Supplied with voltage V_{in}, the voltage drop V_{VD} over R_{SR} is given by

$$V_{VD} = \frac{R_{SR}}{R_{PTC} + R_{SR}} V_{in} \tag{4.1}$$

with PTC resistance R_{PTC} and shunt resistance R_{SR}. Assuming R_{SR} to be constant, the voltage divider approach allows for calculating R_{PTC} by reading the voltage between PTC and R_{SR}. Voltage reading is done with V_{in} supplied. Therefore, heating and reading are decoupled. R_{SR} also draws power and is thus heated. Not only is overall power consumption increased but also R_{SR} must have a temperature independent resistance to avoid additional measurements or feedback within the voltage divider. To minimize these effects, R_{SR} is chosen small compared to R_{PTC}. This reduces power consumption and consequently heating of R_{SR}, decreasing waste energy and requirements of the R_{SR}, namely the temperature independence of its resistance. The first point is important not only for efficiency but also to allow the sensor's surrounding to keep its temperature and not be heated by waste energy of the sensor. With R_{SR} sufficiently small, i.e. $R_{SR} \ll R_{PTC}$, equation 4.1 simplifies to

$$V_{VD} \approx \frac{R_{SR}}{R_{PTC}} V_{in} \propto \frac{1}{R_{PTC}}. \tag{4.2}$$

The last step assumes constant V_{in} and R_{SR}. Any change in thermal surrounding alters the temperature of the PTC and thus its resistance. As this is inversely proportional to V_{VD}, the thermal surrounding can be quantified by reading the voltage V_{VD} over the shunt resistor R_{SR}. Therefore, signal of the sensor/pixel refers to V_{VD} from here on. For an areal sensor multiple of those voltage dividers are connected in parallel and to the same power supply unit (PSU). Disregarding small variances due to the fabrication process, all PTCs in the circuit are provided the same voltage and thus same heating power and temperature behavior. A circuit schematic is given in figure 4.4. From here on the circuit of contacts, voltage dividers, and heating elements as a whole is denoted 'sensor' or 'device' and the individual voltage dividers are denoted 'pixels'. For simple readout of the individual pixels, a contact line from the voltage divider (at V_{VD}) is printed to a connection area where all readout lines can be accessed with a standard ribbon cable. Via this cable the sensor is connected to some readout electronics to process the data. For demonstration purpose an `Arduino-compatible Mega 2560` is used. Its 16 ADC channels read the analogue voltage of the pixels one by one and send digital data to a connected computer. The code used both on the Arduino and computer are shown in

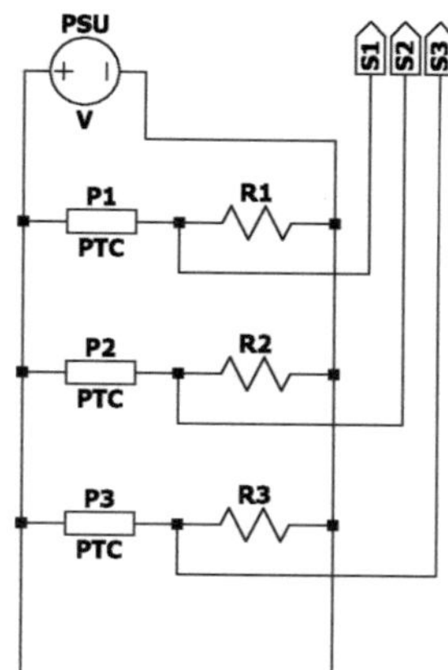

Figure 4.4.: Schematic of the circuit of three pixels of the level sensor. The PTCs PX form with the R_{SR}s RX a voltage divider. The voltages are fed to the readout with contacts SX.

appendix A.1. Although bulky it suffices as proof of concept. For future prototypes a dedicated readout electronics may be assembled or even printed.

While the above describes the general concept, two explicit applications were tested. As they differ from each other in requirements, tailor-made layouts are realized. One sensor is intended for level sensing (chapter 5), its design optimizes spacial resolution and usage. The second sensor is intended for airflow sensing (chapter 6) with a design more general. The latter also has an additional element in the circuitry: each pixel contains a current limiter CL. It is placed between the voltage divider and the readout connection. Thus, PTC, R_{SR}, and CL share one node. The design is depicted in figure 6.3. CL has the purpose to protect the readout from current spikes in case of PTC malfunction and high voltage to the readout. For that its resistance R_{CL} is high though the exact value is not of importance since no current flows through CL. As verified by tests it has no influence on the voltage reading of the corresponding pixel. CLs are not present in the level sensor to reduce substrate space at the expense of a less protected readout. Furthermore, the contact lines of the printed elements are longer than required in the case of the airflow sensor. Reason for that is cost and time efficiency in case of later adjustments. The conducting paths can be printed with the same screens and only the screen of the according material has to be reworked.

To avoid unnecessary waste heat generation and concentrate sensing on one point, PTCs are printed with an active area of 1x1 mm^2. Longer channels provide no benefit. Only a small length is heated and accordingly the resistance increases. This in turn results in reduced current over the entire length. Away from the hot spot or hot line transverse to the current flow, resistance is low and thus power draw and temperature are low, too, as described by equation 2.1. As a result, additional length of the PTC simply adds to overall resistance but does not provide further sensing. Previous tests show cold resistance of

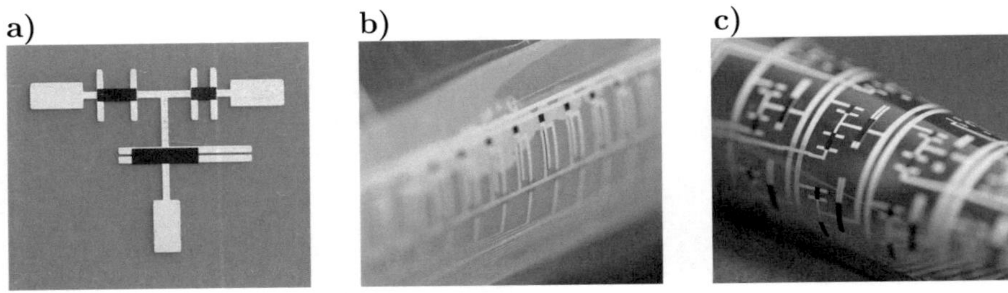

Figure 4.5.: Photographs of pixels in different layouts. The design in a) was used to verify printing and concept. b) shows a section of the level sensor on a PP cup, c) shows the layout of the wind sensor. The level sensor skips the CL and has a different spatial arrangement, otherwise dimensions are the same.

under $10\,\text{k}\Omega$ and hot resistance of over $100\,\text{k}\Omega$[1]. Accordingly, a resistance of $2\,\text{k}\Omega$ for VD is targeted. With a sheet resistance of around $50\,\text{k}\Omega/\square$ dimensions of $w \times l = 5 \times 0.2\,\text{mm}^2$ were chosen for VD[2]. The dimensions of CL were set to $w \times l = 2 \times 1\,\text{mm}^2$.

4.2. Verification

In this section the concept described above is experimentally verified.

4.2.1. Ink Characterization

Electric characterization of exemplary pixels was done on separate single test pixels. They provide the same components of same dimensions as the full devices. Additional contact pads for each connection are printed on the test circuit to allow for a fast prototyping interface for test equipment. Photographs are shown in figure 4.5. To ensure similar properties to the full devices, the test pixels were printed with the devices on the same substrates.

Resistance measurements of individual components were taken with a voltage sweep from $0\,\text{V}$ to $0.1\,\text{V}$. Higher voltages were avoided to avoid unintentional heating of the samples. The PTC element has a resistance of $R_{\text{PTC}} = (8073 \pm 658)\,\Omega$ at $30\,°\text{C}$. Heated externally on a hotplate the resistance increases by a small amount up to a temperature of around $60\,°\text{C}$. With further temperature increase, resistance strongly increases, but only for a short temperature range. Between $62\,°\text{C}$ and $65\,°\text{C}$ the increase is largest. At $70\,°\text{C}$ resistance is at $(149 \pm 44)\,\text{k}\Omega$. From resistances at $30\,°\text{C}$ and $70\,°\text{C}$ a PTC ratio of (18 ± 6) can be calculated. At temperatures exceeding this jump in resistance, resistance slowly decreases with increasing temperature. NTC behavior above T_{op} is observed in literature as well [53, 60]. There

---

[1] Own samples and samples from Christian Willig, InnovationLab GmbH.

[2] Personal communication with Johannes Zimmermann and Benjamin Obermayer, both InnovationLab GmbH.

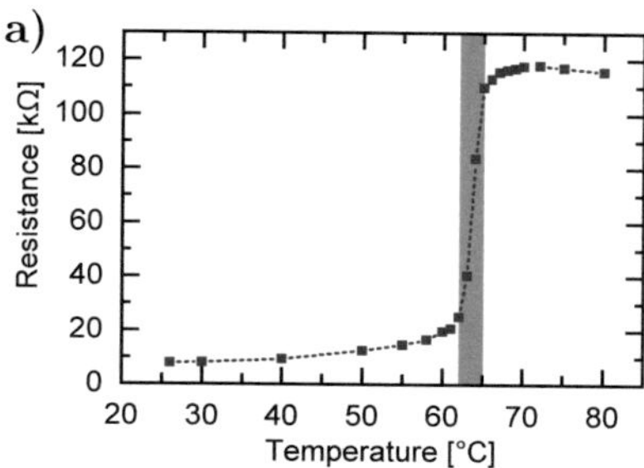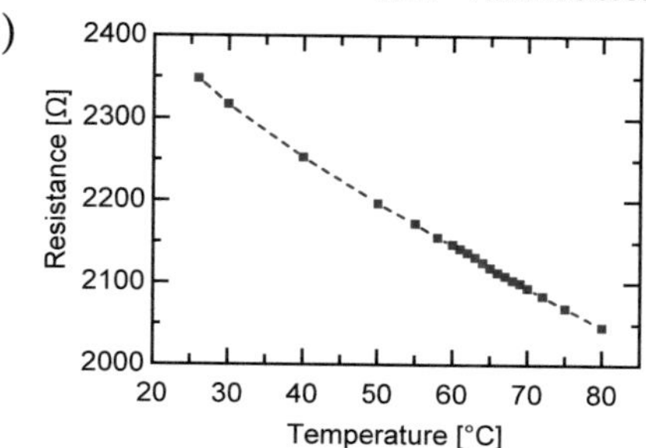

Figure 4.6.: Experimental resistance vs temperature curve of exemplary test structure. a) shows the characteristic for the 1x1 mm^2 PTC element. Highlighted in green is the range between 62 °C and 65 °C. b) depicts the R-T curve for the 5x0.2 mm^2 VD element. The sample was externally heated and probed with 0 V to 0.1 V.

it is explained by lower viscosity in the melted polymer and the polymers conductivity at higher temperatures. The recorded temperature dependent resistance is depicted in figure 4.6a. It resembles the conceptual depiction in figure 4.3a. Therefore, it is expected to work as intended for the concept described above.

R_{SR} at room temperature has a resistance of $(2188 \pm 99)\,\Omega$. Compared to the PTC element the temperature behavior is inconspicuous. Upon heating, its resistance decreases almost linear with temperature, see figure 4.6b. Until 70 °C resistance dropped about 10% to $(1992 \pm 79)\,\Omega$. In the device its lower resistance compared to the PTC greatly reduces power draw to a fraction of the PTC's. The resulting heating of the R_{SR} is minimal and R_{SR} is not expected to have a significant temperature increase during operation. Due to R_{SR}'s neglectable temperature change and the resulting neglectable resistance change - which would only be 10% at 70 °C - R_{SR} can be assumed constant.

CL is also not expected to change temperature as it is intended as reading line and thus not carrying any current. At 30 °C it has a resistance of $(13.2 \pm 0.9)\,\text{M}\Omega$. Its characteristic follows R_{SR}, having a almost linear decrease with temperature. At 70 °C it is still at $(10.7 \pm 0.9)\,\text{M}\Omega$. The relative change is thus 20%. However, it serves as a current limiter without interfering with the voltage. Therefore, the exact resistance is of little importance. In case V_{VD} exceeds the maximum readout voltage, CL can be utilized in a second voltage divider prior to the readout electronics to break down the voltage to a suitable level. Further tests down the line render the voltage levels compatible and no second voltage divider is required.

Reaching T_{op} requires some energy and doesn't happen instantly. The exact time depends on various factors, mostly the ability of the environment to take heat from the sensor. Two examples of heating period are displayed in figure 4.7. For a sample suspended in air, which results in little heat transfer to its surroundings, equilibrium is reached after about 3 s. Higher heat transfer increases the heating period as more of the power is transferred

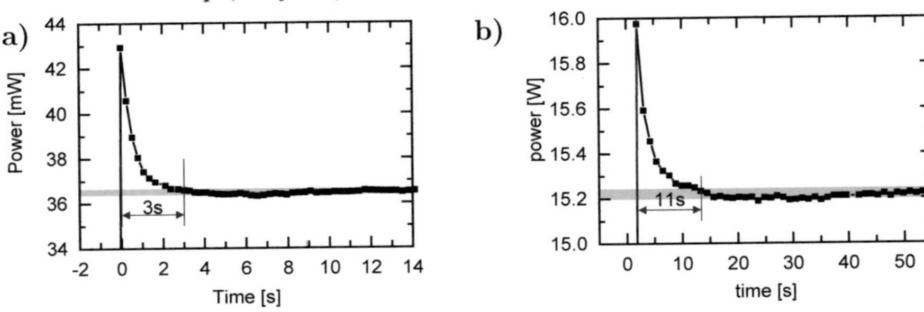

Figure 4.7.: Power consumption during heating period. In the first few seconds power consumption is higher as the temperature of the PTC rises. As it is heated it reaches its operation temperature and power consumption decreases. Time greatly varies between samples in air, thermally isolated, (a) and on a metal plate, thermally well connected (b).

from the sample instead of used for temperature increase. The test sample on a metal plate requires 11 s to settle on power draw and temperature. Heating time is also influenced by the supply voltage and according maximum power used. Voltage also influences sensitivity of the sensor and should be evaluated depending on the application. In summary, it takes the sensor a few (≤ 10) seconds to be ready.

The theory of the origin of the PTC effect and quantification thereof is still ongoing research [42, 53]. Especially reproducibility of the resistance levels after crossing the critical temperature T_{op} is challenging [100]. Choosing a commercial material mitigates this problem and the cycling stability should be satisfying. This was experimentally tested. Voltage supply of a test PTC sample was switched on and off, maintaining the status for around 70 s each time. After switching voltage 100 times no major effect on the current levels was detected (figure 4.8a). The material can therefore be assumed to be stable in prolonged cycling.

Figure 4.8b shows I-V curves of a PTC sample in different thermal surroundings. For low voltages the resistance is nearly constant and independent of the environment. This is due to negligible heating of the sample. It generates not enough heat for heat flux to get important. Thus, it behaves as an ohmic resistor. With increasing voltage however, an increase in resistance can be observed and it becomes linear with voltage. The increased voltage and thus power draw of the PTC results in heating of the sample. An increase in temperature results in increased resistance, see figures 4.6a and 4.3, especially around T_{op}. Effectively the self-limiting heater effect kicks in. For a set voltage above ohmic behavior the resistance is higher for the sample suspended in air compared to the sample on a metal plate. This is because the ability to conduct thermal energy is different for air and metal. Air is less thermally conductive. Thus, the sample suspended in air requires less power to be heated to its operation temperature T_{op}. Any additional heating by further increased voltage gets restricted by the strong temperature-dependent

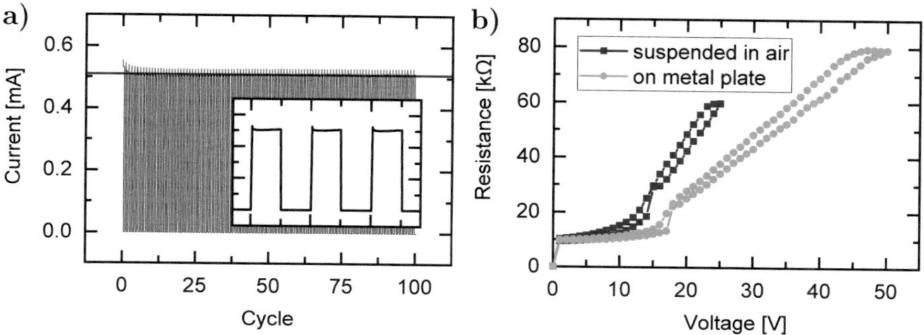

Figure 4.8.: Switching stability of (a) and influence of surrounding (b) on PTC. a) The sample was subjected to 30 V for 70 s before being idle at 0 V for 70 s. The red line highlights the constant current level at 30 V independent of switching number. Three cycles are shown in the inset. The spike in the beginning of each switch is due to a cold sample and thus low resistance at room temperature. b) The sample was heated by applying a voltage. It was either suspended in air (dark) and thus thermally well isolated or on a metal plate with good thermal conductivity (light).

resistance around T_{op}. Metal transfers more thermal energy than air and thus the sample in contact with a metal plate requires more power to reach T_{op} to counter the losses by heat transfer. For that reason it has a lower resistance than the air sample at same voltage.

4.2.2. Verification of Test Devices

It is recommended to choose the voltage of the sensor based on application estimation, sensor dimensions, and/or calibration tests. For the given sensor a supply voltage V_{in} of 30 V was chosen. (R_{SR} is designed to have minimal influence on the PTC's heating and is therefore ignored for this decision.) On a metal surface 30 V is already enough to induce a strong change in resistance. In air the sample can withstand 30 V without burning and getting destroyed. With the values recorded for R_{PTC} and R_{SR} the signal V_{VD} is expected to vary between 23 % (at room temperature) and 1 % (at 70 °C) of the input voltage V_{in}, compare equation 4.1. Given a supply voltage of $V_{\mathrm{in}} = 30$ V, the signal of sensor pixels reaches almost 7 V. However, the chosen microcontroller board for readout electronics is limited in its input to an absolute maximum of 5.5 V [107]. Since the signal voltage V_{VD} can exceed the maximum input of the readout, the supply voltage V_{in} is ramped up to preheat the sensor and increase PTC resistance before reaching 30 V. With $V_{\mathrm{in}} = 23$ V and R_{PTC} at room temperature level, V_{VD} would be at 5.29 V and thus within specifications, whereas V_{PTC} would be at 17.71 V. Even on a metal plate R_{PTC} increases at that voltage to twofold the resistance of room temperature, see figure 4.8b. Therefore, PTC heating sets in early enough when slowly increasing voltage to 30 V. For higher voltages (required for applications in cold environments or with a strong heat sink), the CL has to be used in conjunction with another resistor for a second voltage divider to protect the readout board. As this was not the case in this work, the only precaution to protect the readout was a voltage ramp at start-up of the device.

The influence of CL on the signal was tested for completions sake. Voltage levels of the sensor with active voltage supply were measured on both sides of CL. No difference was detected.

Finally, the sensors were tested to detect various changes in its environment. Individual components are characterized and theory explained, function of the final structure and device needs yet to be verified. For that, several quick attempts to trigger a signal were tested. The most basic and natural trigger is the sample being touched by a gloved finger. Contact with the finger will allow heat to be transferred from the sensor. Thermal conductivity of the glove made of nitrile rubber is a factor of 3 to over 10 higher than that of air, depending on the source [108–111]. Also, behind the glove is effectively a liquid/blood cooled finger, transporting away transferred heat. Expectedly, touching the sensor cools the sensor and increases its signal. Exemplary signal curves are depicted in figure 4.9a. Sensitivity towards different triggers was rudimentary tested. In a still room, calmly waving a hand over the sample causes enough of a draft to temporarily cool the sensor and trigger a visible peak in the signal (figure 4.9b). Although the peak is in the order of the noise of the raw signal (ca. $\pm$ 0.01 in relative signal), an averaged curve enhances peak visibility. Consequently, the device works as intended with a high sensitivity. Interestingly, in the same experiment, three peaks are visible despite only two waving motions. The third peak at around 145 s is believed to arise from air movement caused by opening and closing a door almost 4 m away from the sensor. Otherwise the laboratory was still during this experiment, including the testing engineer. This high sensitivity towards small air movements allows for applications in security as motion detection or intruder alert[3]. Sensitivity is further tested against other influences. The laser (class 2, wavelength: 532 nm, optical power: <1 mW) of a handheld presenting device (Kensington) is pointed directly onto the PTC elements of a test sample. As the laser introduces a new heat source with unchanged heat transfer from the sensor, the PTC is expected to heat up and with increasing resistance, the signal goes down, as described in equation 4.2. Data confirms, that the laser is enough for a signal change to be detected, see figure 4.9c. Interestingly, signal oscillations can be seen when recording with high sensitivity. An example is shown in figure 4.9d. It shows the signal of all 16 pixels of one sensor. Fast Fourier transform of the data indicate an oscillation frequency of 10 Hz. These oscillations are synchronous over all pixels of the sensor. Therefore, they are believed to arise from external factors like interference during signal processing. Since the sensitivity is high enough to detect a laser pointer and small air movements despite the oscillations, no attempts were undertaken to reduce or identify them.

---

[3]Thanks to Prof Uwe Bunz for this idea.

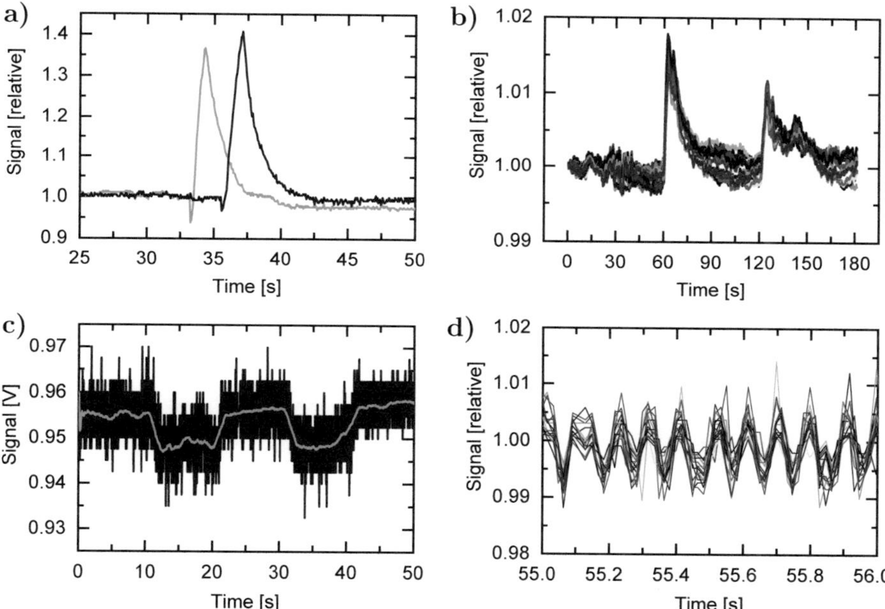

Figure 4.9.: Signal of function tests. a) Response of two pixels to touch of a gloved finger. b) Signal while waving over the sample. Two waves were performed, one after 1 min and the other after 2 min. All 16 pixels of the test device react similar. Shown is the moving average of the signal. c) Sensitivity towards a laser pointer. A handheld laser pointer (green, <1 mW) is pointed on a pixel in 10 s intervals. The raw signal is colored blue and the moving average red. d) Oscillations in raw signal. All 16 pixels of a device are shown to demonstrate the synchronous oscillations.

These tests validate the function of the sensor. Several arbitrary changes in (thermal) environment or sensor heating/cooling cause a change in the sensor's signal. The potential of the sensor for many different applications is thus high. Various methods of adding to or removing power/heat from the sensor can be detected. Two applications are tested and demonstrated in the following two chapters.

5. Level Sensor

Liquid level sensing is essential for handling process liquids and thus heavily required in industry and research [112–116]. Most level sensors are designed to be inside the container with contact to the liquid [114–123]. Aggressive liquids can thus not be used or required special treatment of the sensor. Replacing the sensor in case of faults is hindered as it requires work inside the container. Depending on the liquid, build-up on the sensor has to be considered [124]. Here, a liquid level sensor based on the concept of chapter 4 is introduced for use outside of the container. This level sensor is also described in [91].

5.1. Sensor Principle and Design

The sensor principle is based on heat transfer. Quantity and speed of transferred heat depends on materials around the sensor. Speed or heat per time unit transferred are quantified by thermal conductivity λ. Amount of heat absorbed by temperature increase of the material is expressed by specific heat capacity c. Both λ and c are material specific properties. Heat flux into material thus also varies between materials. Therefore, the sensor concept introduced in the prior chapter qualifies to differentiate between them.

To provide utmost range of possible applications, the liquid height is sensed from the outside through the container. This provides easy installation and integration in existing containers and structures. It also circumvents compatibility issues with the liquid, e.g. in case of acids or solvents. As a result, the heat generated in the PTC element must travel through the container wall into the liquid inside. To provide sufficient and differentiable heat flux, thermal conductivity through the container wall must not be too small. This usually results in a limited thickness of the container wall.

For best results the sensor layout is designed specifically for this application. Multiple dimensions are usually not required for simple liquid level sensing as gravity makes for flat surfaces. Only the height matters, horizontal resolution is unnecessary. Therefore, the pixel array is arranged in a 1D line. To increase vertical resolution the SR is at the same height as the PTC. For best compromise of overall observable height and resolution, a spacing of 5 mm between pixels was chosen. Readout and power supply are on the same short side of the rectangular substrate to simplify connections and electrical accessibility. The design is illustrated in figure 5.1. Dimensions of PTC and VD are the same as described in chapter 4 at 1x1 mm^2 and 5x0.2 mm^2, respectively. These active elements overlap the silver contact lines by 1 mm on either side, thus they look 2 mm longer than their electric active size. CL is skipped in this design; the additional safety of a CL is dispensable for

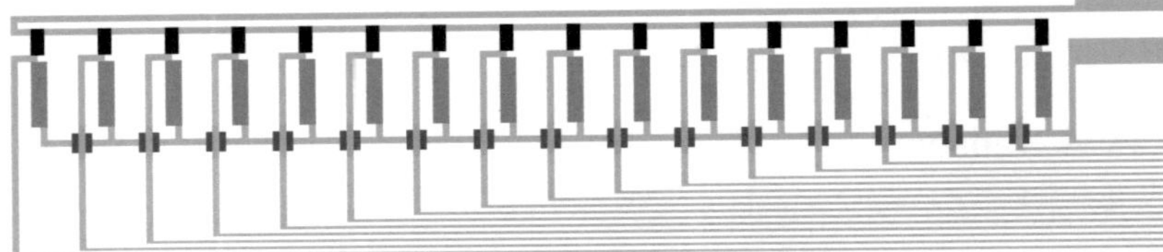

Figure 5.1.: Layout of the level sensor. Connection lines are shown on the right and will be either the top or bottom of the sensor. The lines represent silver contact lines, dark gray rectangles are the PTC elements, gray rectangles the shunt resistances, gray squares the dielectric with conductive silver lines bridging over the dielectric. The PTC pixels are placed every 5 mm, total width of the sensor (top to bottom in this depiction) is 2 cm.

increased compactness. However, care should be taken to protect the readout electronics. The readout electronics in this work - the Arduino-compatible board - proved itself robust against small overvoltage and short current bursts. Nonetheless, the supply voltage should be ramped up instead of suddenly switched on, as advised earlier. The sensor is taped on the outside of a container. For proof of concept a 150 ml polypropylene (PP) beaker with a wall thickness of 0.6 mm was used (figure 5.2a). A small amount of thermal paste was applied directly onto the PTC elements to increase thermal contact. It was verified that the direct contact between thermal paste and PTC doesn't interfere with the electrical properties of the later. Since the property to observe is inside the container, the outside should have as little as possible influence on the signal reading. Therefore, heat flux away from the container must be minimized and constant. With the sensor's active area pointing to the inside of the container, the substrate offers some thermal resistance against the surrounding. A thick (albeit still flexible to fit the shape of the container) substrate with low thermal conductivity is of advantage. For this reason, substrates of PEN were chosen in the following demonstration. It has a thermal conductivity λ of 0.19 W/(m·K) [125] and a thickness of 150 µm. The resulting thermal conductivity through the substrate is 1.27 mW/(mm^2·K). In case of odd surfaces or strong curvature the superior flexibility of PI foil can be advantageous (figure 5.2b). The sensor was also printed on PI foil of 25 µm. PI has a thermal conductivity of around 0.128 W/(m·K) [126] and the heat flux through the substrate is thus 5.12 mW/(mm^2·K). As this is around fourfold of the PEN substrate, it is more prone to external factors like irregular wind around the container. However, under normal conditions with little wind around the container, the levels are still traceable. Further insulation can be achieved by applying thermal insulators or by avoiding extensive airflow around the sensor.

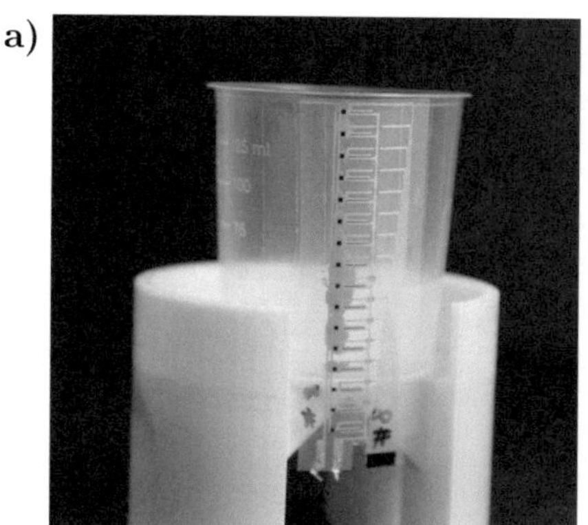 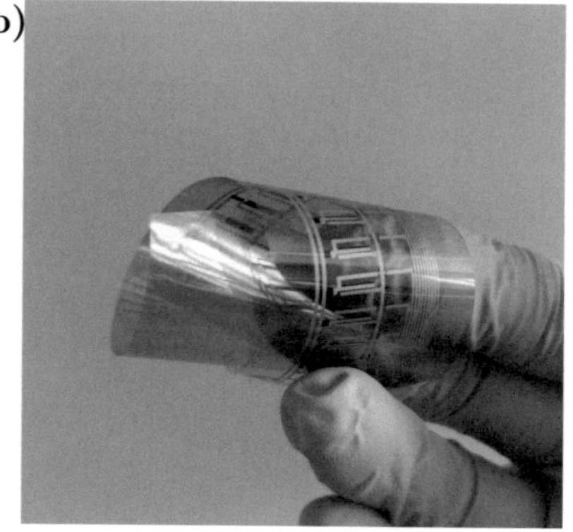

Figure 5.2.: Printed level sensor. a) The level sensor with PEN substrate is applied to a PP beaker and ready for use. Grey thermal paste is between substrate and container. The black squares are the active PTC material. b) The level sensor, here on PI substrate, is highly deformable and thus adaptable to unusual and non-flat surfaces.

5.2. Single Level Sensing

Due to the printing process there are small variances from pixel to pixel. This can be seen in the characterization and the derivation range of single values (see chapter 4), especially the 'hot' resistance of the PTC varies between pixels. The signals of a demonstrator level sensor range from 1.5 V to 1.77 V on an empty container. Thus, a calibration of the sensor was done before measurements. The signals of the sensor on the empty beaker under constant voltage were recorded and used as reference point 'empty cup'. During measurement the obtained signals are output as values relative to the recorded reference point. It eliminates variations originating from the printing process, thermal paste layer and thermal contact to the beaker, and other variations. For increased resolution/differentiation additional points can be added to the calibration. Allowing easy handling and on-site calibration, only one reference point is tried in the beginning.

The sensors capabilities to track a water surface were tested. Tap water was left in a separate container for sufficient time to ensure room temperature. After calibration, the sensor signal was recorded while slowly filling the cup with water. Because of the calibration, all pixels had a signal of 1 in the beginning. Pixels above the water surface stayed at a stable signal of 1. During filling, the signal of pixels increased after water hit the according height. They settled at a relative signal of 1.33 ± 0.02 (figure 5.3). Exception of this were the lowest three pixels in this particular experiment, denoted 1-3 in figure 5.3. They extend below the beaker, see figure 5.2, and serve to detect crosstalk. Pixel 3 changed slightly in signal, 1 & 2 were unchanged. Noticeably the upper two pixels 15 & 16 exhibit a significantly larger error than the other pixel.

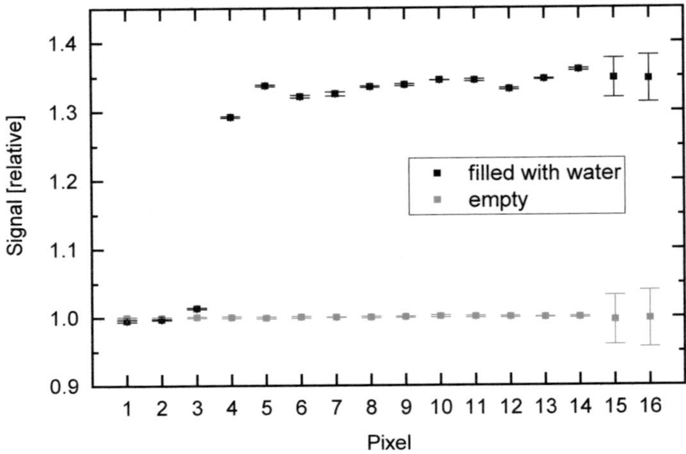

Figure 5.3.: Relative signal of the level sensor. Pixels go vertically from bottom (pixel 1) to top (pixel 16). Pixels 1-3 are below the container bottom. Values depicted are average of 10 s, derivation arises from fluctuations and noise. Pixels 15 & 16 had connection issues at the connector.

Thermal conductivity and specific heat capacity both is higher for water than air (table 5.1). Water is thus expected to be a stronger heat sink for the sensor. Therefore, the sensor takes more power to maintain its operation temperature T_{op}. This is achieved by a lower resistance R_{PTC} which results in higher voltage, see equation 4.1. Pixels 4-16 match the expectation derived from theory. Since pixels 1 & 2 stayed constant, interferences between pixel readings can be excluded. Heat transfer along the substrate can therefore also be neglected. The signal of pixel 3 changed by less than 1.5%. Compared to the 33% of the pixels at water height it is insignificant. Pixels 15 & 16 were found to have had a bad contact at the readout connector. Despite the noise from this bad connection the average value was still usable and matching the expected signal. In figure 5.4 the temporal response of the pixels' signals on the water is visible. At 30 s pixels 4 to 8 significantly deviate from the calibrated value 1 with pixel 9 just starting to show a quick increase. 6 pixels at 5 mm spacing correlate to a water surface at 3 cm from the bottom. This was verified by real time data viewing and simultaneously observing the experiment. To test the influence of the sensor on the water temperature, the following rough estimate yields robust results. It is assumed that all power is used to heat 125 ml of water with no losses. Considering equation 2.1, heating power $P = dQ/dt$, and specific heat capacity $c = dQ/(m \cdot dT)$, the heating rate is

$$\frac{dT}{dt} = \frac{V^2}{R \cdot m \cdot c}.$$

(5.1)

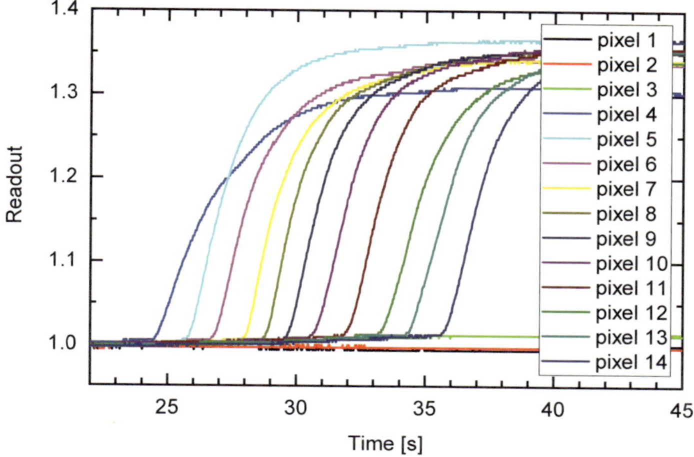

Figure 5.4.: Relative signal of the level sensor with time. Pixels go vertically from bottom (pixel 1) to top (pixel 16). Pixels 1-3 are below the container bottom. Pixels 15 & 16 had connection issues at the connector and are excluded from the graph. As raw data is presented the discrete levels of the ADC of the readout are visible as micro steps.

Table 5.1.: Thermal conductivity λ and specific heat capacity c of relevant materials.

Material	λ [W m^{-1} K^{-2}]		c [J g^{-1} K^{-1}]	
air	0.026	[111]	1.01	[127]
glass (borosilicate)	1.2	[128]	0.83	[128]
olive oil	0.17	[129]	1.97	[130]
polypropylene	0.22	[65]	1.92	[131]
water	0.6	[129]	4.19	[130]

Estimating worst case, R is set to $10\,\mathrm{k\Omega}$ ($< R_{\mathrm{SR}} + R_{\mathrm{PTC,cold}}$). Highly overestimated heating rate for the water in the filled PP cup were $0.17\,\mathrm{mK/s}$. As this is overestimated worst case (no losses, all power into water, PTC not self-limiting), heating of the water can safely be neglected. The experiment provides evidence of the feasibility of using this sensor as water surface detector.

The influence of the liquid container on the sensor function was tested. For that, the same experiment as above was repeated with a (borosilicate) glass beaker with a wall thickness of $3\,\mathrm{mm}$. Apart from container choice all parameters were unchanged. For simultaneous experiments a fresh sensor was used for the glass experiment. Due to the changed frame conditions, a new calibration point was recorded. The reference point for the glass beaker sits at $(1.88 \pm 0.23)\,\mathrm{V}$, which is on average $1\,\mathrm{V}$ higher than for the PP cup with $(0.82 \pm 0.02)\,\mathrm{V}$. A deviation over 10% in voltage between pixels in case of

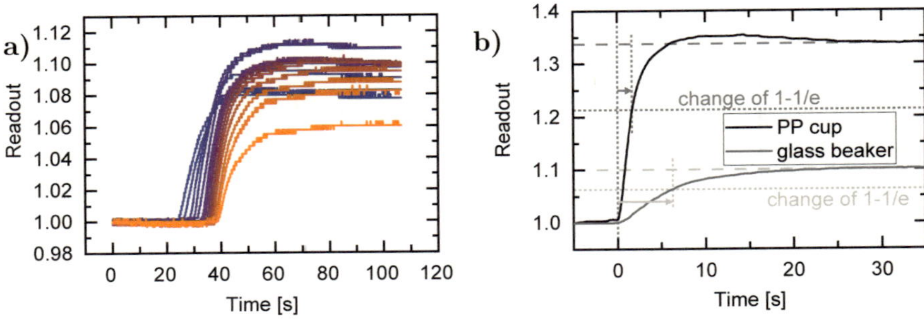

Figure 5.5.: a) Signal over time of sensor on glass beaker. Water is slowly poured into the beaker over ca. 15 s. Pixels are colored from blue for bottom to orange for top. b) Exemplary evolution of signal with time for different containers. At $t = 0$ s water reached the height of these pixels. Indicated are the equilibrium levels and the level and time required to reach $1 - 1/e$ of that level. Curves are Savitzky-Golay smoothed.

an empty glass beaker signifies the importance of proper calibration. Upon pouring water into the glass beaker, the sensor behaved similar as on the PP cup: Until the water surface reaches the height of the pixel, the signal stayed at the calibration value 1. Once the water surface surpassed the pixel, its signal increased. Compared to the PP cup, the relative signal of water at equilibrium is reduced to 1.09 ± 0.01 for the glass beaker. Time evolution of the signals is shown in 5.5a. Depicted in figure 5.5b are the signal evolutions of exemplary pixels on the PP cup and the glass beaker being surpassed by the water surface. The time to reach 1-1/e of its final value increased from (1.7 ± 0.1) s on the PP cup to (6.3 ± 0.4) s on the glass beaker. With around 4 times faster speed and over 3 times stronger signal, the sensor works better on the thin PP cup. 0.6 mm of PP with a thermal conductivity of $0.23 \, \text{W}/(\text{m·K})$ (see table 5.1) account to an energy flow through the wall of $383 \, \text{W}/(\text{m}^2\text{·K})$. This is in a similar range for the glass beaker, with 3 mm and $1.2 \, \text{W}/(\text{m·K})$ it has a theoretical energy flow of $400 \, \text{W}/(\text{m}^2\text{·K})$. However, due to both higher thickness and conductivity, lateral heat flow is greater in the case of the glass beaker. It results in a larger heated area and accordingly less located measuring spots. Thus, point like sensitivity and signal strength are reduced. Additionally, with thicker walls and similar volumetric heat capacities, thermal inertia is higher for the glass container. As a result, response speed is reduced. Despite these drawbacks, the water surface can still be tracked inside the glass beaker, albeit less pronounced and delayed. All in all, the sensor works both on a PP cup and on a glass beaker.

5.3. Mixture Investigation

After successfully detecting and tracking the water-air surface, the level sensor was tested to identify the interface between different liquids. Thermal properties of liquids are typically more similar towards another than to air. For that reason the expected difference in signal of the sensor for different liquids is smaller than in the previous experiment. Therefore, the PP cup was chosen for the following experiment because of its superior signal strength. Additionally, the calibration of the sensor was extended by a second reference point. Conversion of raw signal S_{raw} to calibrated signal S_{cal} with reference values R_1 and R_2 follows the normalization equation

$$S_{\text{cal}} = \frac{S_{\text{raw}} - R_1}{R_2 - R_1} .$$

$$(5.2)$$

Unless otherwise noted, reference points are the empty cup (R_1, $S_{\text{cal}} = 0$) and the cup filled with tap water at room temperature (R_2, $S_{\text{cal}} = 1$). Assuming liquids with thermal properties between those of air and water, the output signal is between 0 and 1.

Liquids chosen to be distinguished are tap water and olive oil. ('Oil' and 'olive oil' are used synonymously in this chapter.) Water is hydrophilic and lipophobic, oil is hydrophobic and lipophilic. Consequently, they don't mix well and separate into pure phases. Due to the lower density of olive oil (0.917 kg/m^3 at $20\,^{\circ}\text{C}$ [130]) compared to water (0.998 kg/m^3 at $21\,^{\circ}\text{C}$ [132]), it swims on top. Because of those reasons, water and oil inside a container are normally not intermixed and the sensor's capabilities to distinguish can be tested. Prior to measurement, the calibration values with the cup empty and filled with tap water were recorded. Then, the cup is filled with $75\,\text{ml}$ of tap water, followed by $50\,\text{ml}$ of olive oil. Data and a photograph of the filled cup are shown in figure 5.6. All pixels started with a value of 0.11 ± 0.02. Upon filling water into the cup, the signals of the pixels at corresponding height (4-10) rose as seen before. Although poured quicker than before, the order of the pixels from bottom to top is visible in the data by their time of reaction. Filling olive oil on top of the water, further pixels (11-14) rose in signal. The slower pouring rate (to avoid spillage) reflects in a larger time delay between signal change of adjacent pixels. Nonetheless, the reaction was similar to water being used. As soon as the liquid surface reached the height of a pixel, the corresponding signal increased rapidly. Pixels at oil height but close to the oil-water interface (11, 12) showed high immediate signal followed by a decrease. The pixel closest underneath the oil surface (14) climbed to a less strong signal but did not decrease afterwards. With time, the signals of all pixels at oil height (11-14) settled around a common value of 0.82 ± 0.06. One pixel just above the oil surface (15) climbed slowly to a value of 0.43. The 7 pixels at water height (4-10) reached a final value of 1.11 ± 0.03. As the oil was poured slowly but not particularly careful, water got displaced and washed up at the container wall. This caused the pixels closest to but above the previous water surface (11, 12) to be cooled by water. Consequently, the thermal flow from the PTC indicated water level. With the interface gradually becoming a horizontal plane, the separation became clearer in the sensor reading. After separation, two main groups of readings formed, one

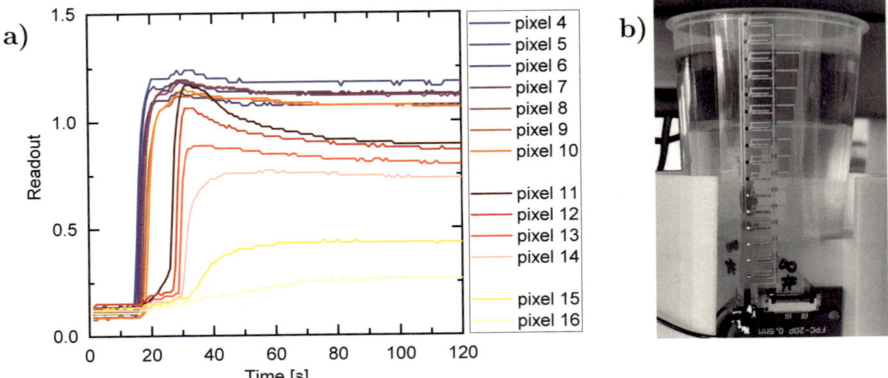

Figure 5.6.: Determining water-oil interface. a) Signal over time upon filling the cup first with water then with olive oil. Not shown are pixels 1-3 not in contact with the cup. b) Photo of the cup filled with water and oil. It shows the state at $t = 120\,$s in a. Pixels are numbered from the bottom (1) to the top (16). The cup is placed on a 3D printed stand to allow for contacting underneath the bottom.

for water and one for oil. Because of oil's lower thermal conductivity and heat capacity compared to water (table 5.1), it provides less cooling to the PTC. Therefore, it requires less power to maintain T_{op} and the signal is lower. However, oil is sill a better heat sink than air, resulting in a higher signal than air-adjacent pixels. As shown in figure 5.6b, pixel 15 was just above the oil surface. Oil creeping up provided some cooling to this pixel, although at a smaller scale. For that reason the signal $S = 0.43$ of pixel 15 is between oil and air. It nicely indicates the precise height of the oil surface at its own height.

If repeated with water and oil poured into the cup simultaneously instead of successively, the result is the same. Data is shown in figure 5.7. Upon filling the cup with a water/oil mixture the signals for all pixels jumped from 0.1 to (1.0 ± 0.1). However, over the next two minutes the signal of some pixels (12-14) decreased, finally settling at (0.78 ± 0.03) whereas the others (5-10) settled after just one minute at (1.02 ± 0.03). One pixel (11) reached a final signal of 0.90. Placement of the sensor pixels is in agreement with these results; the pixels with a signal of around 1 are at the height of water – matching the predefined calibration point of '1' – , the pixels with a signal of around 0.8 are adjacent to oil, and one pixel sits at the interface and yields the intermediate signal of 0.9. Repeating the experiment multiple times always leads to two ranges of pixel signal, one for water and one for oil. The absolute values of the groups vary slightly but can always be grouped in 2 distinguishable ranges. Depending on the total liquid level and sensor placement, the groups are supplemented with another group for air, i.e. below the cup or above the liquids, and pixels with a signal in between the groups.

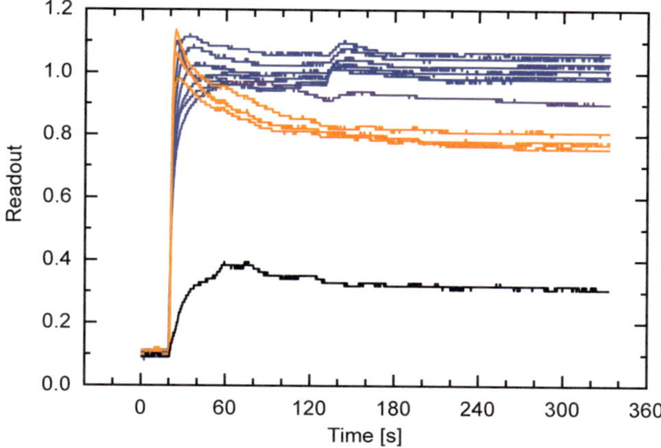

Figure 5.7.: Signal over time upon pouring water and oil simultaneously. Pixels next to water are colored blue, adjacent to oil orange, at the water-oil interface purple and just above the oil surface black. Pixels underneath the cup and further above the oil surface are skipped.

Intermediate signals are very close to a interface or surface, e.g. pixel 15 in figure 5.6 near the oil surface, pixel 11 in figure 5.7 (in purple) at the oil-water interface, and pixel 15 in figure 5.7 (in black) neat the oil surface. In summary, the sensor successfully located both interfaces water-oil and oil-air.

To further challenge the sensor, segregation of a premixed oil-water mixture was examined. As explained above and evident from the previously described experiments, oil and water like to separate after mixing. In some applications it may be of interest to observe the segregation in real time, e.g. in oil production [133] or waste-water treatment [134, 135]. To achieve slow demixing of a mixture, a mixing agent, namely washing-up liquid, is introduced. Thanks to its amphiphilic ingredients, the mixture components can be temporarily mixed to form a uniform blend [136]. With time however water and oil separate again. This separation is to be investigated with the level sensor. To do so, the experiments from before were repeated, but with a thoroughly mixed oil-water-washing-up-liquid mix. Washing-up liquid was used as a mixing agent and thus only in small amounts, just enough to ensure good mixing. The pre-mixed fluid was poured into the PP cup. For increased sensitivity, the mixture was taken as calibration point '1' instead of water while calibration point '0' remained unchanged. The length of the active sensing part exceeded the height of the liquid. Thus, not all pixels could be calibrated with the calibration point '1'. Not fully calibrated pixels did not contribute to the results as they didn't read liquids. They were ignored for the rest of this experiment. Recordings of all pixels can be found in figure 5.8. Due to the calibration, the pixels' signals rose to 1 after filling the cup. Immediately thereafter, the signal decreased for all pixels. Then, pixel by pixel, some signals stopped decreasing and started rising again, the first after ca. 45 seconds. This rise occurred for the pixels from the bottom of the cup upwards and for 6 out of 11 (relevant) pixels. The last to change from de- to increasing did so after 6 min. After 15 min, the signals were separated into two groups. The lower six pixels were at 1.00 ± 0.02, the upper five pixels were at 0.76 ± 0.02.

With progressing time, the liquid at the bottom changed from opaque to translucent, eventually forming a clear horizontal interface between an translucent phase at the bottom and an opaque phase at the top. After formation, the interface moved upwards for about 5 mm. The translucent phase was water falling out of the emulsion. Due to its higher density, it accumulated at the bottom. As already established with the pure phases, water yields a higher signal than oil. Water falling out to form a water-enriched phase at the bottom could thus be seen in the signal of the sensor: the higher the water content, the higher the signal and the more translucent the liquid. The top layer contained more oil and thus decreased cooling capacities, resulting in a decreasing signal. Notably, as the interface moves past pixel 9 after 6 min, the reading of said pixel changes from decreasing - indicating oil-enriched - to increasing - indicating water-enriched phase. Therefore, our sensor distinguishes between the two liquid phases and yields real time observation of the segregation.

5.4. Outlook

In the aforementioned sections the level sensor was successfully employed to detect the water surface in a PP cup and glass beaker as well as to observe separation of an oil water mixture. Due to the flexible nature of potential substrates (figure 5.2) the sensor can accommodate unusual containers. Depending on the hardware connection between sensor and readout it may be advisable to move the pins of the sensor to the long side. Because of the curvature of the cup and the stiffness of the flatband cable connector there were some mechanical connection problems. This can also be overcome by further adjusting the design of the sensor and integrating/positioning the readout lines accordingly. Thanks to screen printing the design can simply be adjusted to suit many different applications. The layout can be adapted to suit different spacial resolution, thermal properties of containers, etc. For ease of installation the thermal paste can be screen printed directly on the sensor [137]. This would also ensure a cleaner look, less material waste, and more similar thermal environment for each pixel.

Heat transfer along the container wall proved to weaken signal strength and response speed, but heat transfer through the container wall is essential for the sensor to work. Consequently, thin container walls are generally recommended. Materials with non-uniform thermal characteristics could also prove well suited. For example, thermal conductivity in quartz depends on the crystal direction; in (001) direction it is double of that in (010) direction

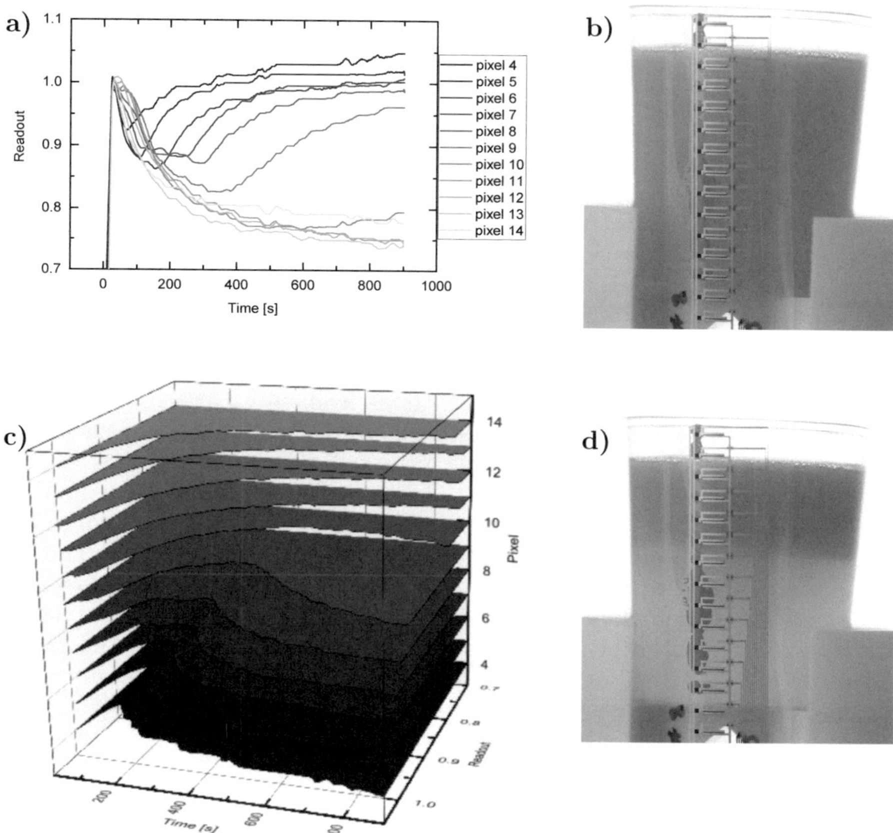

Figure 5.8.: Segregation of water and oil. a) shows the readout over time. Pixels are colored by height from bottom (dark) to top (light). Data is smoothed for better visuals. b) is a photograph of the stirred and filled mixture. c) depicts a 3D representation of the data from a. d) is a photograph of the same mixture from b after ca. 10 min. Photographs and data are from separate experiments but the same sensor.

[138]. Hence, a quartz window in a liquid container does not only provide view inside but also allows for ideal thermal characteristics for the level sensor.

Another method of increasing sensitivity of the sensor is to put it directly inside the liquid itself. Although not the goal of this work, a quick experiment to test the sensor's capabilities was done. Placing the sensor inside the container greatly improves thermal contact and thus heat transfer to the liquids as both sides of the sensor are only separated by a thin substrate from the liquids. It circumvents the challenge of finding a suitable container for sufficient heat flux albeit setting new requirements to the sensor itself and its encapsulation. Without encapsulation, parts of the sensor might just wash off or create a electric short, depending on the liquid. To prevent this, the sensor was laminated. Lamination typically involves high (for polymers) temperatures, adhesive over the entire face, or both.The required temperatures and adhesives have negative to destroying impact on the sensor; none of the fully laminated sensors worked. As it was not the goal of this work to establish an encapsulated sensor for internal use, the lamination options were not exhaustively tested and a proof of concept deemed sufficient.

For proof of concept the sensor was encapsulated by gluing a bare substrate on top of the sensor with adhesive only on the edges, prohibiting contact between glue and printed electronics. 467MP from 3M was used as adhesive due to its tight seal and water resistance [139]. Such a sealed sensor was held vertically inside a container and calibrated against air. Thereafter, the container was filled with tap water. The sensor's reading are visualized in figure 5.9a. Because of the electric connection not being submersible the sensor is turned upside down compared to the sensors glued on the outside of the containers. For that reason pixel 1 is at the topmost and pixel 16 is at the bottommost position. As the sensor has much better contact with the liquid, signal strength for water relative to air is also stronger with 2.33 ± 0.12, a factor of 4 higher than for the PP cup. Figure 5.9b visualizes the difference in signal strength and response speed.

Two problems occurred while testing the encapsulated sensor inside the container. Inside the encapsulation resided some air providing thermal insulation and reducing thermal contact with the surrounding. Also oil tends to adhere to the sensor substrate. If emptied or refilled with different liquids, the readings are off due to the thin oil film on the sensor. Both problems can likely be overcome with the right choice of substrate and lamination/encapsulating technique. Trapped air in the laminated sensor allows for further interesting results. The sensor was placed inside an empty container with an air bubble on the bottom of the sensor which was not initially seen. Upon filling the container with water, the bubble inside the sensor got squeezed, was detected by the observing scientist and slowly crept upwards. As it was passing the PTC elements and temporarily cutting off contact to the surrounding, the bubble can be seen in the sensors signals, depicted in figure 5.10.

For a more robust application a thorough calibration is advised. In the experiments presented here two reference points at most are enough to

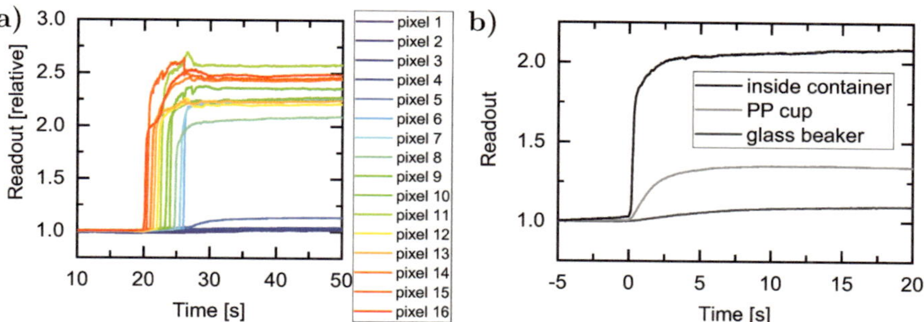

Figure 5.9.: Readout of laminated level sensor inside container. a) At $t = 20\,$s the container is filled with water. Pixel 16 is at the bottom of the container, pixel 1 at the top. To avoid water on the electric connection, the container is not fully filled. b) Signal strength of a sensor on the outside of a glass beaker, on the outside of a PP cup, and of a laminated level sensor inside the liquid. The data from the laminated sensor is of pixel 8 in a.

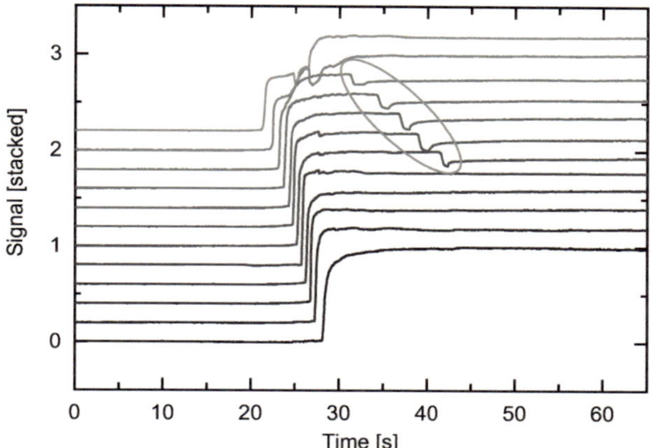

Figure 5.10.: Signal of air bubble in laminated sensor. The sensor was placed in a beaker which was then filled with water. An air bubble was inside the level sensor at the bottom and slowly moved upwards, resulting in a dip in signal, highlighted green. To better distinguish the signals of the sensors, they are shifted along the y-axis. Pixels are colored from light, bottom to dark, top.

distinguish liquids and identify interfaces/surfaces. With changing liquid (or surrounding) temperature the calibration has to be adjusted. Due to the external nature of the sensor, calibration with reference setups is possible without disassembling any equipment or stopping any procedure. It shall however be noted that differences in heat flux can still be detected with the sensor even in non stable environments.

All in all, the concept presented in chapter 4 was successfully applied as a level sensor. The proposed design proved effective for determining the liquid level inside a container while the sensor itself is on the outside. A water surface gives a distinct step in the sensor's signal allowing for easy determination of the water level. This was proven for different containers. Furthermore, oil and water result in different signals, allowing to also identify the height of the corresponding interface. Additionally, the segregation of an oil-water mixture could be observed in real time. The formation of an interface could be tracked with the sensor's reading. This proves the suitability of the sensor to observe not only liquid levels inside a container but also to track segregation or mixture of liquids.

6. Airflow Sensor

While the sensor concept introduced in Chapter 4 works well as a level sensor, it is not the only possible application. In the following, the use as wind sensor - particularly surface airflow - is described.

It aims to support aerodynamic wind tunnel experiments by mapping the surface wind speed over test objects. This offers an alternative to pressure taps often integrated into wind tunnel models to map pressure distribution [140, 141].

Another possible application is the detection of (deep) stall in aircraft. If the angle of attack for an airfoil is increased too far, airflow separates from the upper part of the airfoil and lift is reduced [142]. In case of the whole aircraft changing its angle to the relative air movement, it may became susceptible for deep stall where the aircraft's horizontal stabilizer and elevator are in the wake of the main wing ('deep stall') [143]. This is illustrated in figure 6.1. The aircraft looses its ability to change the angle which most often leads to unrecoverable loss of control and crash [144]. Therefore, it is important to detect stall of the aircraft's wing. While there are techniques and devices to do so nowadays [145](e.g. Safe Flight Instrument, LLC, NY, USA has several stall warning products) this can also be achieved by measuring the airflow over the wing.

6.1. Sensor Principle and Design

The basic sensing principle follows the same as described in chapter 4. The active PTC element is supplied with constant voltage and heats up. Upon reaching thermal equilibrium electrical power input equals thermal flux out of the sensor. Here, however, variations in thermal flux are not determined by surroundings with different thermal properties but by airflow.

Due to electric heating, the PTC's temperature is higher than that of its surroundings. This results in heat flow out of the PTC. The components of the heat flow are, like in the case of the level sensor, thermal radiation, convection, and conduction. Thermal radiating losses account for a maximum of $0.74\,\mathrm{mW/mm^2}$, see equation 2.5. As incoming thermal radiation also heats the PTC, radiation losses are reduced depending on the ambient temperature, e.g. to $0.62\,\mathrm{mW/mm^2}$ at -60 °C or $0.32\,\mathrm{mW/mm^2}$ at 20 °C. Conduction losses depend on the material the sensor is applied on. With the sensor suspended in air and in no direct contact with material, conducting losses are limited to the substrate and conductive paths supplying electricity. In direct contact with material, conduction depends on these materials, as utilized in chapter 5. Nonetheless, in this application the sensor is to be either affixed on a

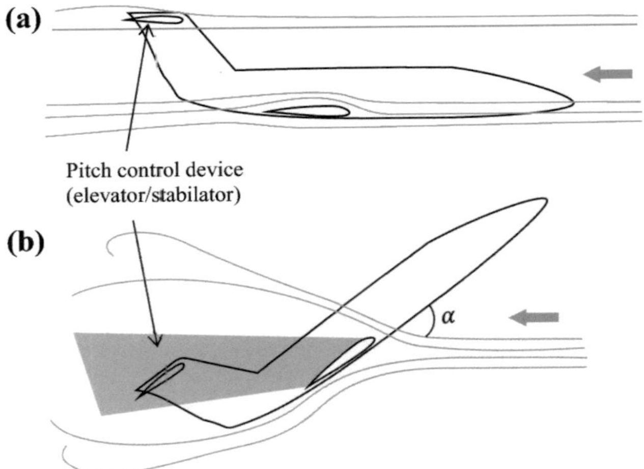

Figure 6.1.: Deep stall illustration. a) Under normal operation, the pitch control device is streamed in by airflow. b) With high pitch, the pitch control device is in the wake of the main wing. Copied from [146] under CC BY 4.0 and modified.

surface or fully suspended in air. Thus, the solid material in contact with the sensor doesn't change. Therefore, thermal conduction can be assumed to be constant and an offset to the variable heat flux. To keep this offset small, materials with low thermal conductivity will presumably work best. Convection on the other hand depends on the airflow around the PTC and is not assumed to be constant. It is formulated in equation 2.7. The heat transfer coefficient h for free convection ranges from $2.5\,\mathrm{W/(m^2 \cdot K)}$ to $25\,\mathrm{W/(m^2 \cdot K)}$, increasing to $10\,\mathrm{W/(m^2 \cdot K)}$ to $500\,\mathrm{W/(m^2 \cdot K)}$ for forced convection [70]. With PTC temperature $T_{\mathrm{PTC}} = 65\,°\mathrm{C}$ and ambient temperature $T_{\mathrm{am}} = 20\,°\mathrm{C}$ this translates to $(0.1 - 1)\,\mathrm{mW/mm^2}$ and $(0.4 - 20)\,\mathrm{mW/mm^2}$ for natural and forced convection, respectively, giving 2 orders of magnitude for sensing range. The correlation between air speed and convective heat transfer coefficient depends on situational parameters like shape and is still debated in literature [147–149]. Exemplary relations from literature are depicted in figure 6.2. Since the cooling strength and thus the energy to counter it depend on the temperature difference between the media, see equation 2.7, the air temperature has to be taken into account. Experiments described here were performed inside labs and air temperature was sufficiently constant. However, for open-air or outdoors application the use of a temperature sensor is proposed.

The layout for the wind sensor of this work differs from the level sensor. One key aspect is the ability to map wind speeds over a surface requiring a 2D structure. Thus, the sensor array consists of a 4×4 matrix with a pixels spacing of $15\,\mathrm{mm}$, see figure 6.3. Active PTC area is $1\mathrm{x}1\,\mathrm{mm^2}$, shunt resistor area is $5\mathrm{x}0.2\,\mathrm{mm^2}$ and current limiter area is $1\mathrm{x}2\,\mathrm{mm^2}$, contact pads were chosen to be $10\mathrm{x}10\,\mathrm{mm^2}$. The most important dimensions, PTC and shunt

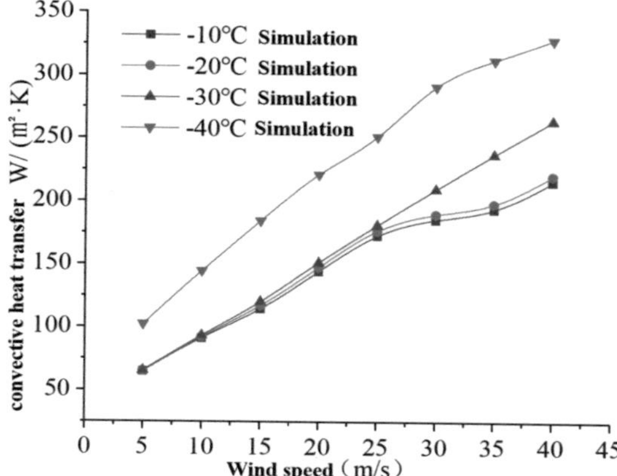

Figure 6.2.: Convection coefficient vs Air speed. Copied from [149] under CC BY 4.0.

resistor geometry, are the same as for the level sensor. The silver contact lines adjacent to the electric components exceed the necessary length. This was done to allow for easy size adjustment of the PTC, SR and CL if required. On the sensing application it has no influence except minimum spacing / resolution. Although one dielectric bridge of the readout lines is not strictly required (of the bottom left pixel in figure 6.3) it was printed for consistency and symmetry. The sensor matrix is read out with the same hardware and software as before with only the visualization changed to 4 x 4 from 16 x 1.

6.2. Experimental Results

Firstly, the correlation between airflow and a single PTC element is investigated. A PTC test sample with width $w = 2.5\,\text{mm}$ and length $l = 1.6\,\text{mm}$ is put centered inside a tube of diameter $d = 80\,\text{mm}$. The substrate is aligned with the airflow, i.e. the substrate's normal vector is perpendicular to the pipes center line. The sample is held in place by crocodile cables which double as electric connection. It is directly connected to a variable constant voltage power supply. Current was measured with a standard multimeter. One side of the horizontally placed pipe is equipped with a 80 mm fan to provide variable wind speed. Airflow is set via the fan's voltage and follows the empirical relation[1]

$$\dot{V}_{\text{air}} = A\sqrt{V_{\text{fan}} - B} \tag{6.1}$$

with parameter A determining the airflow and B the lowest voltage with the fan still spinning [150]. To verify this empiric equation it is fitted to experimental data of a fan from the same manufacturer as the one used.

---

[1]The mathematical approach was chosen due to lack of suitable reference anemometer.

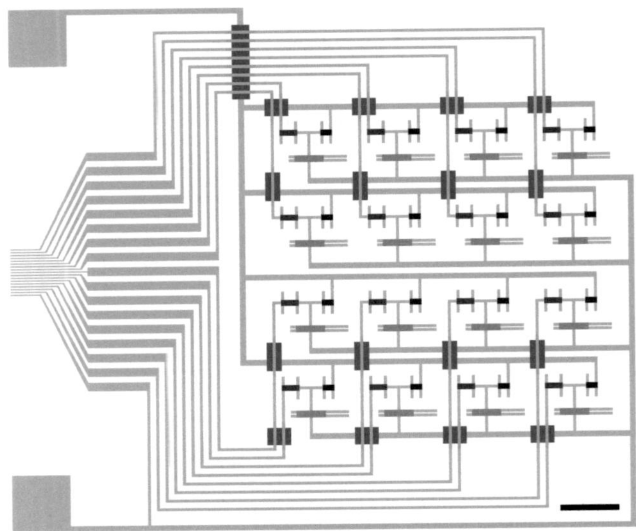

Figure 6.3.: Design of the airflow matrix. The black scale bar of 10 mm serves visualization and is not present in the layout.

Data is taken from [151]. Fitting A and B to the data results in $A = 14.76$ and $B = 3.40$ at an adjusted coefficient of determination R^2 of 0.99. Thus, equation 6.1 is verified. These results are adjusted to the fan used which has a maximum airflow of $35.8\,\mathrm{m^3/h}$ at $12\,\mathrm{V}$ [152] and an experimentally determined minimum voltage of $4.2\,\mathrm{V}$, so equation 6.1 for the given fan yields

$$\dot{V}_{\mathrm{air}} = 12.82\sqrt{V_{\mathrm{fan}} - 4.2}\,. \tag{6.2}$$

With equation 6.2 the voltage of the fan can be translated to airflow and average wind speed within the tube. Thus, PTC power dependency on wind speed can be recorded (figure 6.4). The wind speed of $v_{\mathrm{air}} = 0\,\mathrm{m/s}$ refers to no forced air movement, i.e. $V_{\mathrm{fan}} = 0\,\mathrm{V}$. Natural air movement and convection still occur so the real wind speed is $> 0\,\mathrm{m/s}$. The relation between PTC power and wind speed is almost linear with a slight hysteresis except for no forced air movement. In direction of only increasing wind speed a linear fit results in $P = (5.31 \pm 0.03)\,\mathrm{mW/mm^2} + (1.06 \pm 0.02)\,\frac{\mathrm{mW/mm^2}}{m/s}$ with an adjusted R^2 of 0.997, a fit over the whole circle (except $0\,\mathrm{m/2}$) yields $P = (5.46 \pm 0.06)\,\mathrm{mW/mm^2} + (0.99 \pm 0.04)\,\frac{\mathrm{mW/mm^2}}{m/s}$ with an adjusted R^2 of 0.973. Assuming constant temperature difference ΔT - which is satisfied via the step like PTC behavior of the material used - convection is almost linear in this regime [147, 148]. Towards lower air speed, the convection coefficient decreases more than linear [147]. These results verify the suitability of the approach as a quantitative wind speed sensor.

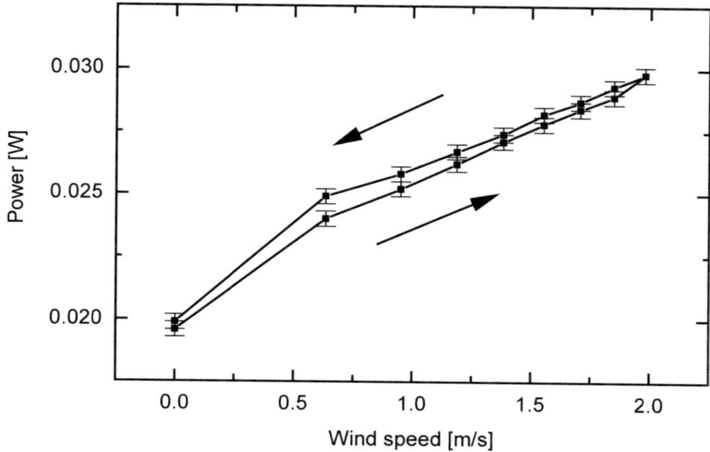

Figure 6.4.: Power draw of single sensor vs wind speed. The sensor with $w = 2.5\,$mm and $l = 1.6\,$mm is supplied with $10\,$V and suspended in the center of a horizontal pipe with a diameter of $80\,$mm.

An airflow of $35.8\,\mathrm{m^3/h}$ through a 'wind tunnel' with a diameter of $80\,$mm equals an air speed of $v \approx 7\,\mathrm{km/h}$. This is around half of $13.9\,\mathrm{km/h}$, the average wind speed in $10\,$m height in Heidelberg where this work was carried out [153]. Therefore, the sensor can resolve air speeds at or below average wind speed, allowing for use in meteorological applications.

Supply voltage depends on thermal surroundings and cannot be specified in general. Fully suspended in air the sensor is relatively isolated and requires only little power to stay at operation temperature T_{op}. Total radiation on both surfaces (front and back) of the PTC depends on ambient temperature T_{am}, e.g. $0.64\,\mathrm{mW/mm^2}$ at $20\,°C$, vide supra. Relative to the total power of the device of minimum almost $5\,\mathrm{W/mm^2}$, see figure 6.4 for $4\,\mathrm{mm^2}$ sample, radiation can be neglected. Convection also occurs on both sides of the substrate. Additionally, the sensor's substrate also heats up and gets cooled by airflow, effectively increasing surface area and thermal inertia. Contact lines and connected cables provide another heat sink, further offsetting the required electric power to keep the sensor at T_{op}. At $2\,\mathrm{m/s}$, the sample of figure 6.4 draws $(29.8 \pm 0.3)\,\mathrm{mW}$. With an area of $4\,\mathrm{mm^2}$ and convection occurring on both sides (double the surface area), the heat transfer coefficient calculates to $(82 \pm 1)\,\mathrm{W/(m^2 \cdot K)}$. This is a factor of 2-4 above literature values of convective heat transfer coefficents, see figure 6.2. For no forced air movement, the heat transfer coefficient is $(55 \pm 1)\,\mathrm{W/(m^2 \cdot K)}$. The difference in heat transfer between no forced air movement and wind of $2\,\mathrm{m/s}$ is around $27\,\mathrm{W/(m^2 \cdot K)}$, more in line with literature, yet still higher. This can be explained by the aforementioned influence of substrate and contact lines and the resulting larger cooling area. With fixed surroundings, a calibration routine accounts for these influences. Without calibration, qualitative results are still well within the capabilities of the sensor, rendering it suitable for a wind change sensor or differential mapping.

With this setup a supply voltage of 10 V is sufficient to obtain results. However, 30 V were also tested successfully. If the sample is in direct contact with a thermally highly conductive material, the thermal flux into the material is high and the required power to feed the PTC is consequently higher. It is then recommended to introduce an isolation layer between sensor and surface. The substrate itself provides some thermal resistance. With thicknesses of 150 µm and 25 µm for PEN and PI, respectively, they allow for $1\,\mathrm{mW/(mm^2{\cdot}K)}$ and $18.4\,\mathrm{mW/(mm^2{\cdot}K)}$. While the first option offers some resistance, the thin PI substrate has too high a conductivity of 828 mW/pixel at $\Delta T = 45\,\mathrm{K}$ and pixel size of $A = 1 \times 1\,\mathrm{mm^2}$. Thus, the thermal properties of the object in contact with the sensor have to be accounted for. Many possible applications use materials with low thermal conductivity, e.g. ABS (acrylonitrile butadiene styrene) $(0.25\,\mathrm{W/(m{\cdot}K)}$ without filler [154]) or similar thermoplastics for 3D printing of wind tunnel test models [140, 141] or composite materials including carbon fiber (roughly up to $5\,\mathrm{W/(m{\cdot}K)}$ [155]) for aircraft wings [156]. Thermal conductivities in that range should pose no major issue for sensitivity. Therefore, the proposed sensor can easily be adopted in those fields.

6.3. Sensor Matrix

A printed matrix is depicted in figure 6.5. Similar to the level sensor, PEN and PI were used as substrates. Electrically, no difference could be found between substrate materials. Mechanically, the PI substrate offers superior mechanical flexibility, adapting to many surface forms. PEN is stiffer, yet still adapts to most bends. For substrates loosely attached to an object, the stiffness of PEN proves superior. It can also be used suspended in air to test airflow at heights. The flexible PI substrate buckle under airflow if not tensioned or adhered properly on the object, flapping around if suspended in air without object.

To test the spatial resolution, patterns were drawn on a sensor using a compressed air gun set to low pressure. Prior to that, the sensor was calibrated similarly to the level sensor: Signals are given relative to the sensor in still air lying on plywood, calibration point '1'. Upon blowing air from the compressed air gun at the sensor, the signal of the respective pixels increased, see figure 6.6. Neighboring pixels also show an increase in signal albeit less pronounced. The increase indicates a reduced PTC resistance R_{PTC} which is due to a lower temperature caused by increased/forced convection. With the air stream light and located, individual pixels (in contrast to the whole matrix) can

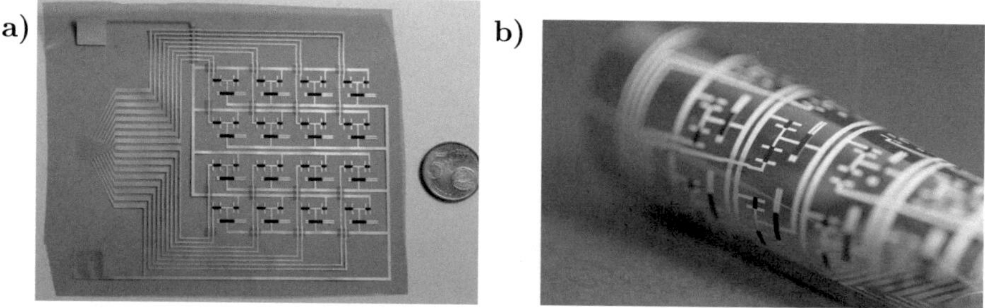

Figure 6.5.: Photos of airflow sensor. a) Full 4x4 matrix on Kapton foil. b) The same matrix bent and still working.

be subjected to strong convection. Because the air cannot penetrate the substrate and is deflected over the surface of the sensor, it also cools pixels in the vicinity. This effect is stronger with higher airflow due to the increased air mass/volume. The targeted pixel however yields the highest signal, clearly differentiated from its neighbors. It can thus be concluded that resolving spatial distribution of airflow over the sensor was successful.

Since the above experiment only tested against airflow pointed perpendicular on the sensor, airflow along the sensor's surface is investigated in the following. The experiment simulates an object in moving air with part of it in the wake of another object. The sensor, lying on plywood, was placed in front of a fan. Between fan and sensor, a solid object blocks half of the sensor from the airflow as illustrated in figure 6.7. In the same figure, the acquired data is show as well as a snapshot of the readout. During fan operation, all pixels show a signal > 1. However, one side is on average more than double above the reference point than the other side. Thus, the main airflow is limited to one side and was successfully detected by the sensor. The barrier doesn't sharply block any air movement in its wake. Rather, it blocks the main movement while turbulent air still moving over the 'blocked' sensor part, providing some cooling and consequently increased signal. To note is the fall time of the signal after the fan was turned off. While a clear drop can already be noted after $< 0.5\,\mathrm{s}$, it takes around $10\,\mathrm{s}$ for the signal to reach the idle value, i.e. calibration point 1 for no active fan.

All in all, the airflow sensor was tested successfully. It clearly resolved the location of an airflow directly on the substrate. Further, regions of different airflow speed over the sensor could be detected. For airplane stall warning however, a signal fall time of $10\,\mathrm{s}$ is too much. An application as such consequently would require further optimization and testing. In the experiments executed in this work, airflow speeds were generally low and in the range of natural winds. Airfoils however are subjected to much higher velocities. This drastically increases cooling for the sensor and results in a much higher signal. Therefore, the author believes the concept is generally

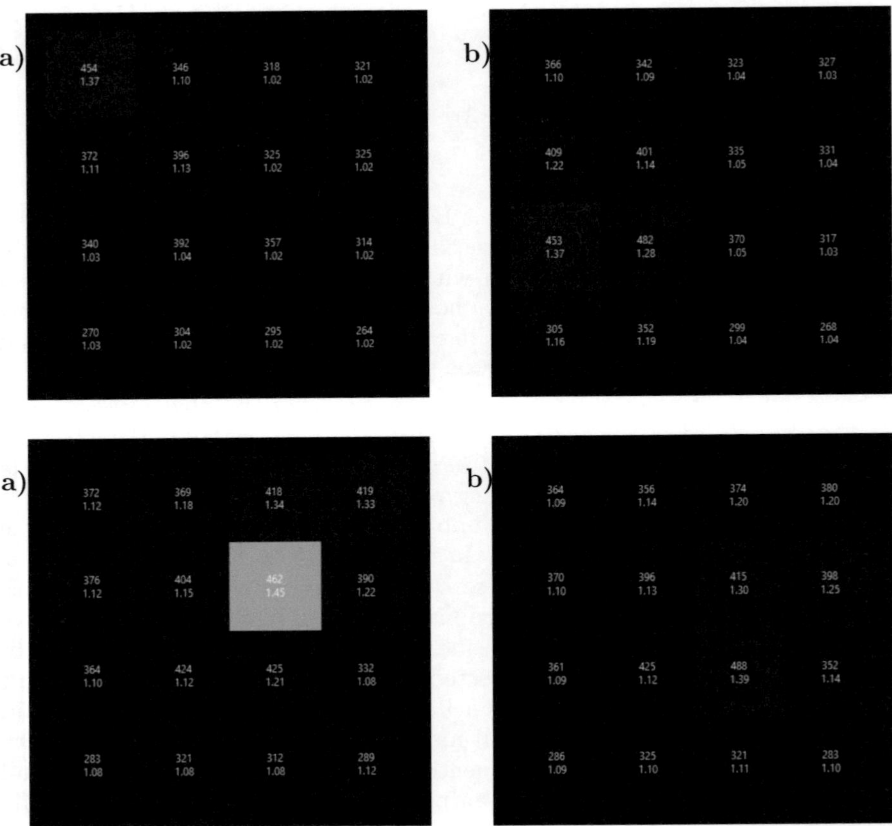

Figure 6.6.: Compressed air gun blowing on airflow sensor. The air gun is pointed at pixel 11 (a), 13 (b), 32 (c), and 33 (d). Color changes from black at 1.0 to blue at 1.4, upon which is switches to cyan. Further details can be taken from the code in the Appendix.

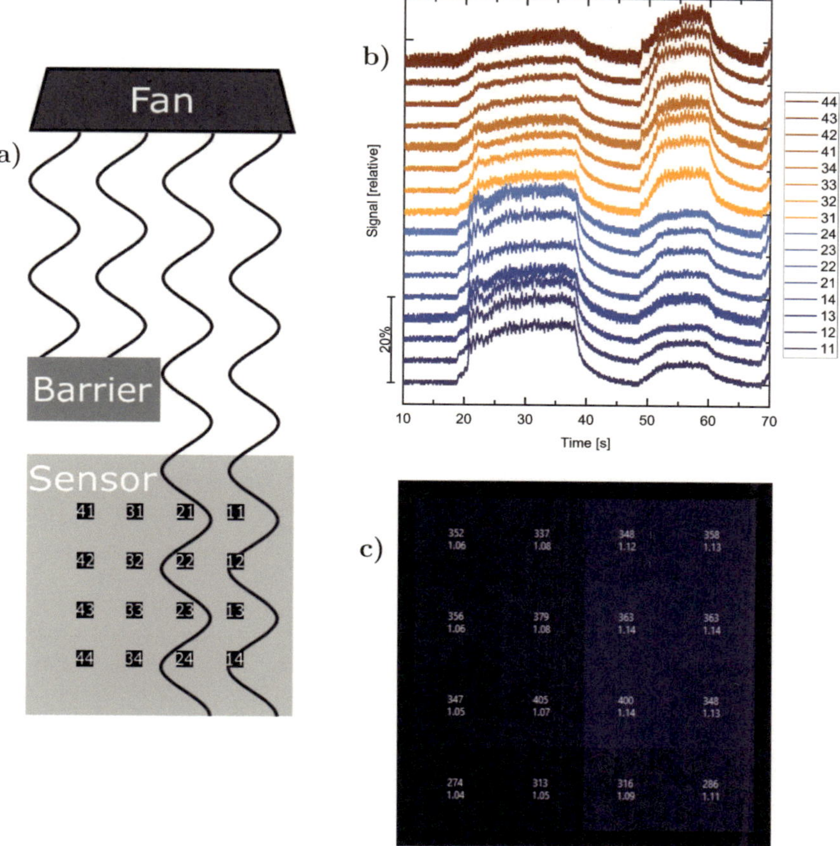

Figure 6.7.: Airflow over half the wind sensor. a) Experimental setup. A fan blows air to the wind sensor, half of the path is blocked by a solid barrier. b) Readings of the sensor pixels. A barrier is placed in front of the right side of the sensor and the fan turned on at 20 s for ca. 20 s. Then, the barrier is moved to the left and the fan turned on again. Figure a represents the setup between 21 s and 38 s. c) Live feed of sensor. The snapshot shows the readout at $t = 27$ s.

suitable. Wind speeds on the other hand were well resolvable. With proper calibration and environment temperature compensation, the sensor - or a single pixel thereof - has the potential as a meteorological wind sensor. Another promising application is the use in aerodynamic testing. Applying the sensor on the surface of a model in a wind tunnel should allow to map the airflow directly on the surface, identifying wakes and and flow separation. Wind tunnels provide a well controlled environment, thus no temperature sensing is required at the sensor to optimize its operation. In summary, the sensor presented here is a promising candidate for meteorological application and evaluation of aerodynamics in wind tunnels.

7. OFET

The sensor as described in chapters 4 - 6 features multiple pixels that can be arranged in a 2D array. Although only 16 pixels are realized, the concept can be scaled to a higher pixel count and larger area. Actuation is done by one readout line per pixel. Because of the required constant heating of each pixel, they have to be permanently connected to the power source. For large pixel counts in 2D arrays a matrix circuit is common however. Goal of such approach is to reduce the number of required connections. A pixel in a passive or active matrix is accessed by one connection for its line and one for its column. Individual pixels are read/actuated by multiplexing both lines and columns. In a 32×32 matrix this reduces the required connections from $32 \cdot 32 = 1024$ to only $32 + 32 = 64$ (plus optional common lines). The difference between active and passive matrix is an additional switching component per pixel in the active matrix. Advantage of the additional element is reduced crosstalk and better isolated signal of individual pixels. One goal of the 2-HORSIONS project is an active sensor matrix. For that, OFETs as switching components inside a temperature sensor matrix are tested.

7.1. Semiconductor Characterization

The OFETs are based on a new semiconductor material XPRD01B06 from BASF. It is investigated on batch-to-batch variations and energy levels.

On each of three production batches 'OP1' to 'OP3' a chemical analysis is performed to review batch-to-batch variations. For each material batch a film was spin coated on a silicon substrate. Film thicknesses were determined to $35 - 40\,\mathrm{nm}$ with Dektak. The spin coated films were investigated with XPS to get insights into the chemical composition. A survey spectrum of a film with batch OP2 is depicted in figure 7.1. This spectrum allows to identify chemical elements present in the sample. XPRD01B06 contains thiophene and DPP groups[1]. Chemical structures of these groups are depicted in figure 7.2. Thus, the material contains carbon (C), sulfur (S), oxygen (O), nitrogen (N) and hydrogen (H). Although it is possible to record spectra of hydrogen, this requires highly specialized equipment and was made possible only recently [157]. Therefore, hydrogen is not investigated in this analysis. The other elements are present in the sample recorded as seen from the survey spectrum. From low binding energies (right hand side in figure 7.1) to high ones (left hand side) the peaks are identified as S2p (165 eV), S2s (230 eV), C1s (285 eV), N1s (400 eV), and O1s (531 eV) [76]. The steps at higher energies are

---

[1]Personal communication with Dr Kai Exner, BASF New Business GmbH.

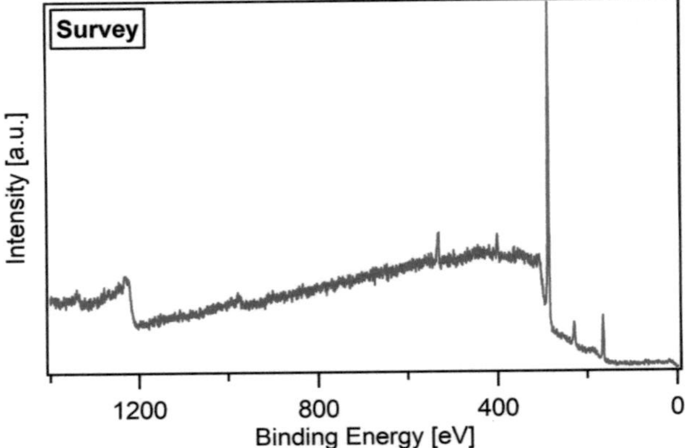

Figure 7.1.: Survey spectra of XPRD01B06 OP2. Peaks over background are identified as, from right to left, S2p (165 eV), S2s (230 eV), C1s (285 eV), N1s (400 eV), O1s (531 eV), Auger lines for oxygen (975 eV, KLL transition), and Auger line for carbon (1220 eV, KVV transition).

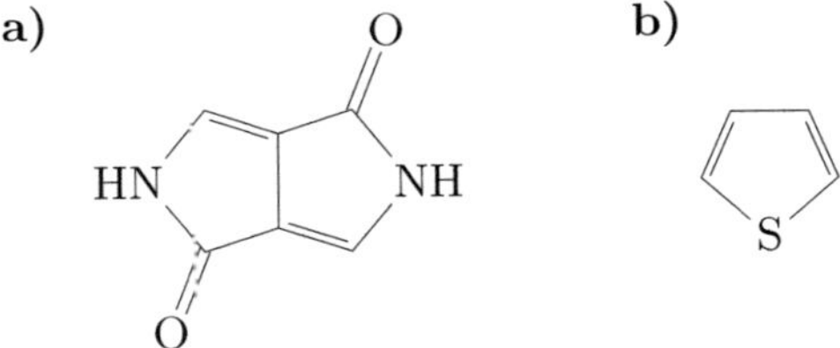

Figure 7.2.: Chemical structure of DPP (a) and thiophene (b).

identified as Auger lines for oxygen (975 eV, KLL transition) and for carbon (1220 eV, KVV transition) [76]. Thus, all expected elements are found while no unwanted/unexpected elements appear in the sample.

As most parts of the spectrum contain no useful information, detailed spectra only of selected regions are recorded. Higher resolved spectra allow for improved quantitative and qualitative analysis. The C1s, N1s, O1s and S2p core spectra are depicted in figure 7.3. Even without quantification of the results, strong similarities between the samples are visible, indicating high reproducibility between batches. After corrections due to setup and method, the N1s orbitals of the samples cause one Voigt profile over a flat background. The S2p graph shows a double peak profile. Background is flat except a small incline at higher binding energies. O1s investigation shows one Voigt profile and a small but broad rising at higher binding energies. Similar observation can be made in the C1s graph. At 285.4 eV is a strong peak, flanked by a second rising at 287.6 eV. Closer analysis and fitting reveals the strong peak to consist of two profiles, one centered at 285.3 eV and one at 285.9 eV. The integrated intensities are listed in table 7.1.

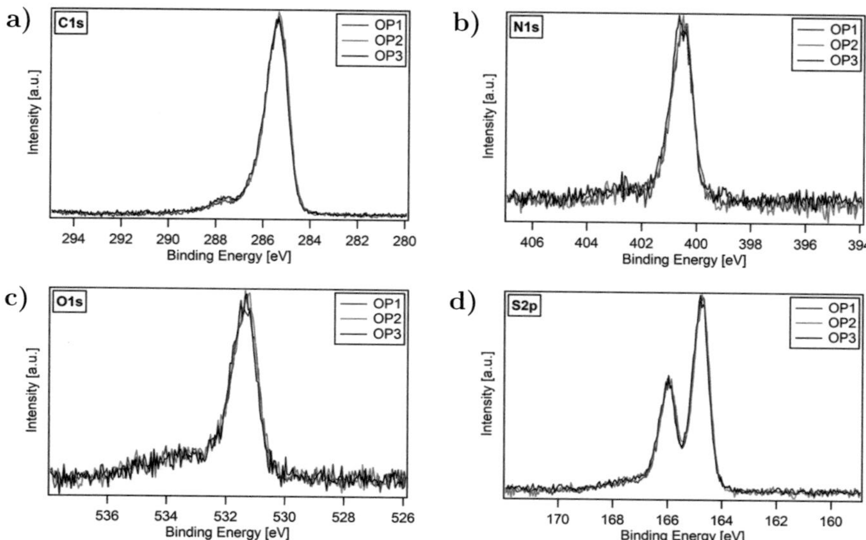

Figure 7.3.: Core level spectra of XPRD01B06. Depicted are the energy regions of the levels of a) C1s, b) N1s, c) O1s, and d) S2p. To show similarities, they lay on top of each other, partly hiding OP1 (and OP2) behind OP3.

Table 7.1.: Integrated intensities of batches OPx of XPRD01B06.

Orbital	OP1	OP2	OP3
C1s	6520	6779	6667
N1s	396	383	464
O1s	601	621	592
S2p	917	937	952

These results allow for conclusions about chemical bindings and structures within the sample. Nitrogen is present only in the DPP monomer and only in one chemical configuration. Thus, all nitrogen atoms are in the same state and have the same binding energy for the N1s orbital, broadened by varying chemical environment and equipment. Similarly, sulfur is present only in the thiophene monomer in one configuration. However, as the orbital under observation is the p orbital it is a doublet with different binding energies [76, 158], S2p3/2 at 164.8 eV and S2p1/2 at 166.0 eV. It is in good agreement with literature: the *handbook* gives a binding energy for atomic sulfur of 164.0 eV for the $2p_{3/2}$ and a split of the doublet of 1.18 eV, with sulfur bound in thiophene in a very similar region [76, 158]. Therefore, it can be deduced that all sulfur is within the thiophene. Oxygen is also expected to be in only one configuration as it should only be present in DPP. However, experiments reveal most O1s electrons to be in the same state at 531.3 eV, but there are also a significant number of electrons with higher binding energies, indicated by the broad rising between 532 eV and 536 eV in figure 7.3c. This was attributed to surface oxidation of the film. As this material is intended to work in a transistor in ambient condition the tested sample was also subjected to ambient (clean room) conditions prior to the experiment. The visibility of the oxidation layer is magnified by the fact that XPS only probes the surface of the film. Water inclusion as a reason for the oxygen is unlikely both by the binding energy ([159, 160]) and the ultra high vacuum (UHV) condition during measurement. Carbon is the main component of the molecule and present in different binding constellations. Thus, it is unsurprising that the spectrum is not made of a single but rather multiple peaks. All in all, the results are pleasing. Components are revealed as expected except of excess oxygen which is attributed to (surface) oxidation.

Integrated intensity of XPS signals can be used to determine atomic ratios. The components should only be analyzed relative to each other and not in an absolute way. The photo ionization cross-sections σ used for equation 3.4 are taken from [84]. Resulting ratios of atomic components are listed in table 7.2. Oxygen to nitrogen is around 1 throughout all samples. In thiophene there is neither oxygen nor nitrogen, one DPP unit has two atoms of both each. The expected ratio of O/N is therefore 1 and in agreement with the experimental results. The aforementioned surface oxidation cannot be seen in these atomic ratios. OP2 contains slightly more oxygen than theory predicts, however, the theoretical value is still within the error range of the experimental value and both OP1 and OP3 show less oxygen than theoretically predicted. The S/N ratio averages to about 2.5, or 5 sulfur atoms per 2 nitrogen atoms. Given the molecular structures this translates to 5 thiophene units per one DPP unit. This finding is consistent for OP1 and OP2. OP3 on the other hand has the lowest O/N, S/N, and C/N ratios, indicating an overall higher nitrogen content. Assuming only a different ratio of thiophene to DPP would suggest 4.5 thiophene units per DPP unit as per S/N $\approx$ 2.25. However, O/N is also lower than the expected 1, originating only from DPP. Therefore, the

Table 7.2.: Atomic ratios of batches OPx of XPRD01B06. For each sample, measurements on different days and spots on the sample were taken and averaged. 'Average' in line four corresponds to the average of all measurements of all batches.

batch	O/N	S/N	C/N
OP1	0.96 ± 0.04	2.46 ± 0.08	29.28 ± 0.66
OP2	1.10 ± 0.15	2.52 ± 0.15	30.91 ± 1.34
OP3	0.83 ± 0.07	2.25 ± 0.06	26.84 ± 1.34
Average	0.96 ± 0.13	2.42 ± 0.14	29.05 ± 1.88

additional nitrogen arises not from more DPP inside the film but from other sources. In conclusion, the smallest repeating building block of XPRD01B06 consists of five thiophene and one DPP unit throughout all batches. Five thiophene and one DPP monomers together attribute to 26 C atoms. With an average of 29 ± 2 C atoms per N atom or 58 ± 4 C atoms per DPP, this leaves 32 ± 4 carbon atoms for the organic side chains whose structure will not be investigated further in the scope of this work.

All in all, the chemical reproducibility between the three batches is good. While OP1 and OP2 show only marginal and negligible variations, the film of OP3 has an increased nitrogen content. As there is also a mismatch in the oxygen to nitrogen ratio, the increased nitrogen content does not originate from a different ratio of thiophene to DPP. The difference however is small and will likely have no influence on device performance. It can thus be concluded that XPRD01B06 has 5 thiophene units per DPP monomer and 32 ± 4 C atoms in both side chains combined.

7.2. Device Characterization

XPRD01B06 was not only investigated as material but also in OFET devices. OFETs were built by project partners from InnovationLab GmbH (Benjamin Obermayer, Dr Stefan Schlisske, Manuel Seifert, Dr Karl-Philipp Strunk) and BASF (Sebastian Raths). The author of this work received the finished OFETs and performed characterization experiments on them.

For that, current voltage (IV) curves were recorded at controlled parameters. A schematic of an OFET is displayed in figure 2.4. The typical IV behavior for a p-type OFET is shown in figure 7.4. For positive gate-source voltages V_G, no charge carriers are injected into the p-channel and no source-drain current I_D can be detected. (Unless otherwise noted, the source contact is grounded, i.e. $V_S = 0\,\mathrm{V}$. The experimental detection limit of the setup used is around $10\,\mathrm{pV}$.) For negative V_G, electrons at the gate contact attract holes from the p-type semiconductor which accumulate near the dielectric and create a p-enriched channel between source and drain. With an applied negative source-drain voltage $V_D < 0$ holes can be injected at the source and extracted at the drain, creating a current $I_D < 0$. The threshold voltage V_{th} - at which the OFET changes from insulating to conducting - is located around

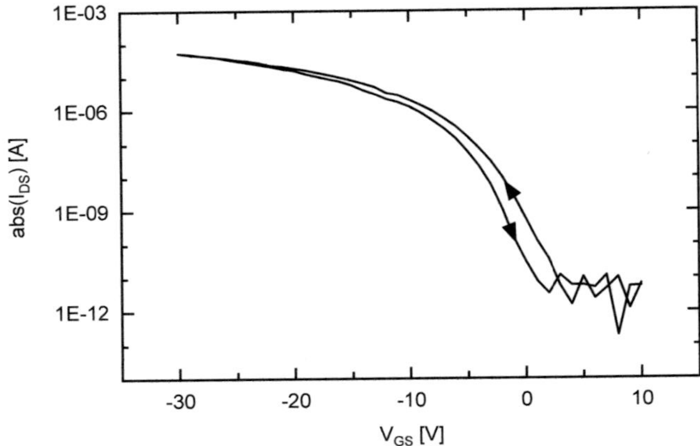

Figure 7.4.: Typical transfer curve of an OFET. Source-drain voltage V_{SD} is set to -30 V.

0 V. Drain current I_D is restricted by the channel resistance which in turn is determined (amongst others) by material, channel width, and channel length. Experimentally recorded resistance consists of the OFET's resistance and additionally the contact resistances. The latter can be significant, especially in organic electronics [161–164]. For that reason, they were investigated.

7.2.1. Contact Resistance

Contact resistance can be determined by the transmission line model (TLM). Underlying is the measurement of the total resistance R_{tot}, consisting of contact resistance $R_{contact}$ and OFET resistance. It is given by

$$R_{tot} = R_{contact} + \frac{L}{W \cdot C_i \cdot \mu \cdot (V_G - V_{th})} \tag{7.1}$$

with channel length L, channel width W, gate-insulator capacitance C_i, and carrier mobility μ [165]. With the denominator set constant by choosing the same parameters, OFET resistance is proportional to L and equation 7.1 simplifies to

$$R_{tot} = R_{contact} + L \cdot X \tag{7.2}$$

with a constant X. Reducing L to zero leaves only the contact resistance $R_{contact}$ for the total resistance R_{tot}. $L = 0\,\text{mm}$ is achieved by extrapolation from several measurements with different lengths but otherwise same parameters. For that OFETs with varying lengths were provided and their total resistance recorded. Data and fits according to equation 7.2 are shown in figure 7.5. The width of these transistor channels is $W = 10\,\text{mm}$. $R_{contact}$ varies with gate voltage V_G and ranges from $(1.3 \pm 18.2)\,\text{k}\Omega$ at $V_G = -30\,\text{V}$ to $(34 \pm 102)\,\text{k}\Omega$

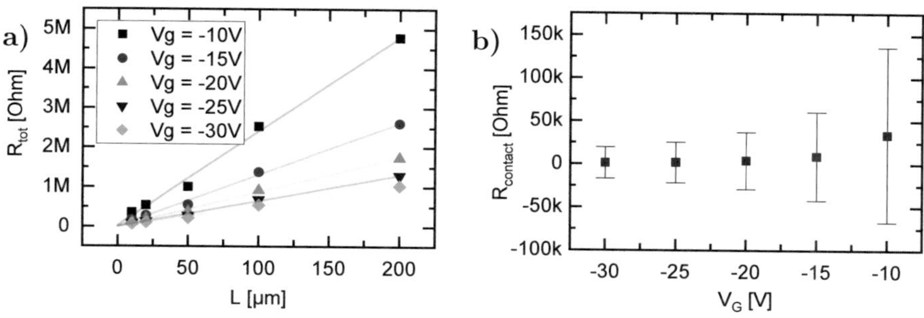

Figure 7.5.: Transmission line model and resulting contact resistance. a) Resistance of devices with different channel lengths. Points are the experimental data, lines depict linear fits for each set of data. Channel width is 10 mm. b) Extrapolated contact resistances. The error arises from the linear fit.

at $V_G = -10$ V. The range of $R_{contact}$ is high and the error from extrapolation is even larger. However, the OFETs are supposed to be switched between $V_G = -30$ V (ON or conducting state) and $V_G > 0$ V (OFF or insulating state) for the intended applications. In OFF state, resistance is supposed to be high and contact resistance doesn't matter. At (relevant geometries and) voltages for the ON state, the contact resistance is $R_{contact} < 20$ kΩ. Usual resistances for the OFETs as used in devices are in the range of megohm. For example, the resistance of the very good OFET of figure 7.4 is $R = {}^{30\,V}\!/_{5.48\times10^{-5}\,A} \approx 547$ kΩ with most OFETs exceeding that resistance. Thus, the main contribution to the total resistance R_{tot} arises from within the channel and the contribution of contact resistance $R_{contact}$ is negligible. For that reason the contacts don't need improvements and their resistance can be ignored in further design developments.

7.2.2. Operational Stress Tests

Organic electronics are prone to degradation [166–169]. The OFETs were thus tested for degradation during operational stress. A fresh batch of OFETs was provided and individual OFETs were put through three different scenarios. Firstly, the high current behavior was tested where the OFET is always on, i.e. $V_S = -30$ V and $V_G = -30$ V. (In this chapter, V_D instead of V_S is grounded for setup reasons. Due to a symmetric layout source and drain are interchangeable and assignments arbitrary.) Secondly, an OFET in always on but blank state was investigated, i.e. $V_S = 0$ V and $V_G = -30$ V. Thirdly, the stability of blank and off OFET was checked, i.e. $V_S = 0$ V and $V_G = 10$ V. The stress tests, labeled 'current bias stress', 'negative bias stress', and 'positive bias stress', respectively, were hold for a minimum of two days. Every 30 min (1 h for current bias stress) the stress was shortly interrupted to record a transfer curve to evaluate threshold voltage V_{th}, on-current I_{on} and charge carrier mobility μ. Results are depicted in figure 7.6a,c,e. Former experiments on similar devices revealed water inclusion over time which was countered by heating the OFETs prior to use[2]. To prevent water to get into the OFET encapsulation was tested. Thus, encapsulated OFETs were also tested for operational stress, see figure 7.6b,d,f. In almost all graphs some kind of oscillation over time is visible. The frequency of the oscillations matches one day. Current bias and negative bias stress tests of both bare and encapsulated OFETs change trends at around 6 o'clock in the morning. This is attributed to a temperature dependence of the device as the lab's temperature did also change during the day. (Temperature dependence is investigated separately, vide infra.)

Current bias stress and negative bias stress yield similar trends. Threshold voltage V_{th} shifts to the negative with time, mobility μ and on-current I_{on} both decrease. As these experiments were carried out on different OFETs, the baseline varies. Nonetheless, the shifts are in similar ranges, less pronounced for negative bias stress. The bare OFET in the negative bias stress behaves similar to the encapsulated OFET in negative bias stress and both bare and encapsulated OFETs in current bias stress, though it has lower I_{on} and μ from the beginning. The shift in voltage over two days of about 2.5 V as well as the relative change in I_{on} ($30\% \pm 5\%$) and μ ($10\% \pm 2\%$ for current bias stress and 22% for the bare OFET in negative bias stress) are of the same order. The initial baseline thus has no major effect on the relative drift. More important is the finality of this burn-in effect. After not being used for some days (and preconditioned with heating prior to the measurement) the OFETs revert to their original baseline, as depicted in figure 7.6a for a downtime of 23 days.

With little surprise positive bias stress has the inverse effect as negative bias stress. In the first few hours V_{th} shifts to more positive values, I_{on} and μ increase. The oscillation over the day is less pronounced as with negative gate voltage. The 11th stress period of the bare OFET was unintentionally stopped after 10.5 min and the OFET was without applied voltage for 103 min (figure 7.6e). In this time period of relaxation the OFET was able to partly recover from previous stress. Afterwards the experiment resumed as intended and the shifts observed before continued. The shifting effect of the parameters is saturated earlier for positive bias stress than for negative bias stress. The encapsulated OFET follows the bare one in behavior in the beginning. However, after prolonged periods of stress the OFET temporarily breaks down, see figure 7.6f. The threshold voltage V_{th} is between 5 and 25 V and on-current I_{on} increases by up to over 25%. This behavior was reproducible with another encapsulated OFET.

To quickly revert the OFET into baseline the effects of reverse voltage were tested. Exemplary one OFET was subjected to positive bias stress and the effects recorded. After 10 times 60 s of positive bias stress (and transfer line recording in between) the voltage was reversed, thus inducing negative bias

---

[2]Personal communication with K.-P. Strunk, InnovationLab GmbH, and S. Raths, BASF.

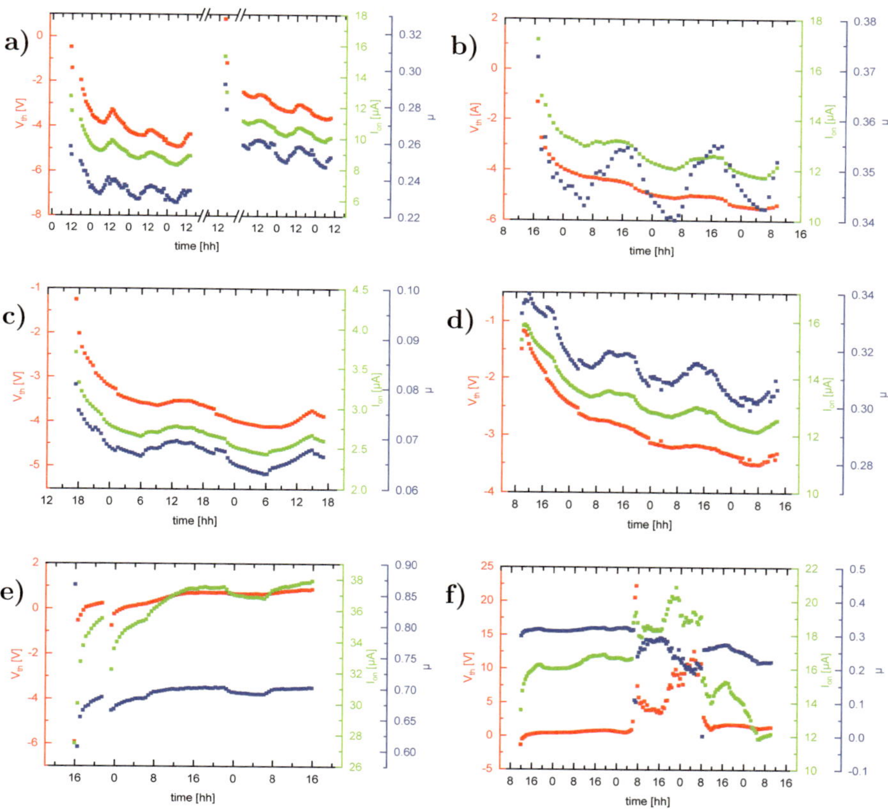

Figure 7.6.: Operational stress tests of OFETs. The time (x axis) is given in time of day. a) shows the current bias stress of a bare OFET. The first skip in the time axis equals 23 days, the second skip is two days. b) shows the current bias stress for an encapsulated OFET. c) and d) show the negative bias stress of a bare and an encapsulated OFET. e) and f) are the positive bias stress graphs for the bare and encapsulated OFETs. Between the 11th and 12th transfer curve in e) (around 10 p.m.) was a time period of 103 min without applied voltage.

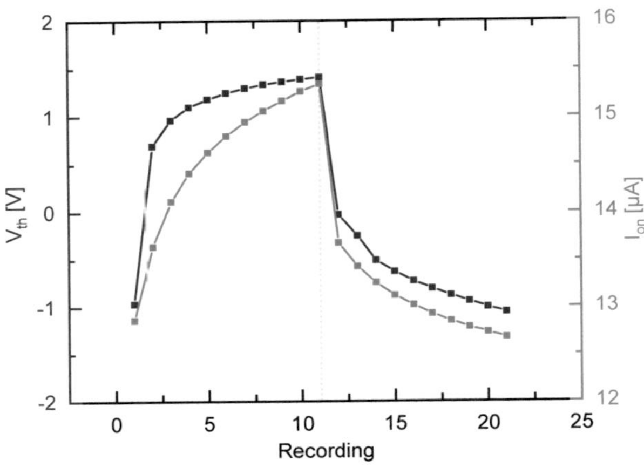

Figure 7.7.: OFET operational stress test and reversed voltage. Each point represents the results of a transfer curve. The OFET experienced $V_G = 30\,V$ (i.e. positive bias stress) for $60\,s$ between measurements in the first half and $V_G = -30\,V$ (i.e. negative bias stress) for $60\,s$ in the second half.

stress. The results are depicted in figure 7.7. Both positive and negative bias stress have the effect as expected and described earlier. As these effects are converse they largely cancel out in the end. Importantly, the shifts in V_{th} and I_{on} are logarithmic with the largest change in the beginning. Therefore, only a short time of reversed polarity can lift most of the shift of prior operational stress.

All in all the threshold voltage V_{th} shifts with operational stress. However, the effect is not permanent and reversible. It also does not affect the on-state ($V_G = -30\,V$) or off-state ($V_G = 10\,V$). The on-current I_{on} shifts as well and has to be accounted for in sensor applications. A reversing operation with e.g. $1\,min$ of positive and $1\,min$ of negative bias stress might be advisable after extended periods to undo these drifts. Prolonged periods (days) of fully switched off OFETs should be avoided as it can temporarily break down during positive bias stress.

7.2.3. Temperature Behavior

Intended application for the OFETs tested here are active matrix backbones for sensing, in particular in conjunction with temperature sensors. Therefore, temperature behavior of the OFETs has to be evaluated. In perspective of the application, OFETs were subjected to temperatures between $20\,°C$ and $90\,°C$. They are heated in $10\,°C$ increments with transfer curves recorded upon reaching target temperature.

In the beginning of the project, the OFETs showed high sensitivity towards temperature. While working as expected at room temperature, at increasing temperatures off-currents I_{off} get too high to be considered off, peaking

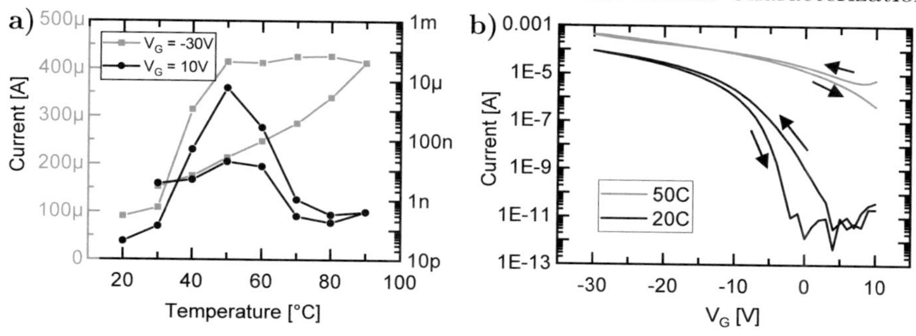

Figure 7.8.: Exemplary OFET current in dependence on temperature. a) shows the on- (light) and off-current (dark), defined as $V_G = -30\,\text{V}$ and $V_G = 10\,\text{V}$, respectively. V_D is set to $-30\,\text{V}$, V_S is grounded. Experiment starts at $T = 20$ °C, increases to $T = 90$ °C and ends at $T = 30$ °C. Respective transfer curves for 20 °C (dark) and 50 °C (light) are shown in b).

at $T = (50 \pm 10)$ °C, see figure 7.8. It shall be noted that I_{off} refers to $V_G = 10\,\text{V}$. This is not necessarily the lowest possible off-current as the threshold voltage V_{th} shifts due to the temperature increase, see figure 7.8b. As a universal and temperature independent integration is desired, the off voltage cannot be adjusted with temperature, thus it is set $V_G = 10\,\text{V}$. On-current I_{on} also increases with temperature. However, it is affected by hysteresis and irregularity, i.e. OFET current cannot be translated to temperature. The on-off ratio diminishes due to high I_{off}, falling below 100 in some cases. For the data shown in figure 7.8 the minimum on-off ratio is 72 at 50 °C. For the application in an active matrix with e.g. 64×64 pixels, leaking current of all off transistors of one line or column would be similar to the current of one transistor in on state. Also the temperature sensitivity in on-current is undesired as the OFET should work robust in many environments and the sensing signal should arise from the sensor in-line with the OFET, rendering different sensing applications possible and independent of temperature. Therefore, these initial batches cannot be used in environments above room temperature.

These results were used by the fabrication team to improve temperature stability of the OFETs. The team adjusted procedures, exchanged process materials, and fabricated new OFETs which were characterized in regard to temperature stability. After some iterations of feedback (also from other sources/experiments) and adjusted fabrication, the OFETs could be stabilized and the on-off ratio is stable over most parts of the temperature range, dropping off towards the high end, see figure 7.9. Even the lowest ratio, located at 90 °C, is comfortably high with over 1000. The resulting signal of individual pixels in an active sensor matrix will therefore be determined mostly by the one OFET in on-state with the sensor in-line. Interference is minimized and the active matrix is expected to be sufficiently optimized for its application. Any further improvement would cost too much effort and time

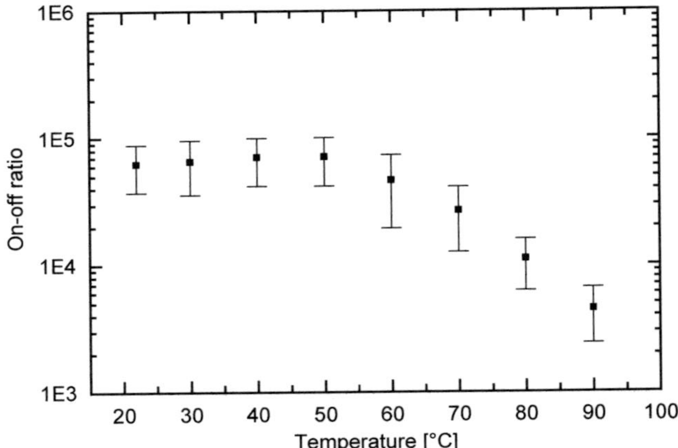

Figure 7.9.: On-off ratio after fabrication optimization. For determining the ratio, off-current is set to 10 pA if recorded value is lower. Values are averaged over a batch of OFETs on one substrate.

for only marginal better results. Therefore, the fabrication team considers their recipe finalized, as is the analysis of the OFETs.

In summary, the OFETs are well suitable for an active matrix backbone for sensing. Resistance originates mostly from the OFET channel itself, rendering any further contact optimization unnecessary. It was shown that extended use or stress on the devices can result in drift of parameters. However, it was also shown that the drift is reversible. Finally, initially strong temperature sensitivity was defeated and the final OFETs are temperature stable. Therefore, the OFETs are feasible for their intended application.

8. Summary and Outlook

The present work is a step towards development of printed sensor matrices. In the frame of the 2-HORISONS project, a new sensor concept was developed and tested for two applications. Furthermore, the newly developed material XPRD01B06 from BASF was analyzed for use in OFETs for an active sensor matrix.

First, the concept of a new sensor was established in chapter 4. It is based on a commercial PTC material. The material is designed for self-limiting heaters. It features a step-like resistance-temperature relation, where the resistance increases sharply at around 65 °C. Due to the relation $P = U^2/R$, the material heats up to around 65 °C and remains at this temperature, as long as constant voltage is applied. Once the material is at temperature, all power going into the sample equals heat flux out of the sample. This was verified by manipulating heat flux with forced convection and thermal conduction through materials with different thermal conductivities. The sensor concept was integrated into an array of multiple sensing pixels. To read the PTC's resistance, a shunt resistor SR was introduced. Together, they form one pixel of the sensor array. The voltage level between PTC and SR is read by the ADC of a microcontroller and subsequently sent to a computer to display and further process the data. Good general sensing capabilities could be verified for the sensor: it was able to detect very slight air movement and the thermal power of a laser pointer.

The sensor concept introduced in chapter 4 was tested in the form of a liquid level sensor in chapter 5. Thanks to its mechanical flexibility, the sensor is able to fit on most liquid containers. The sensor is arranged as a vertical line of 16 pixels with a spacing of 5 mm, which probe their thermal surrounding. Water provides higher thermal mass than air and thus provides more cooling to the sensing pixels than air does. As a result, the resistance of the PTC is lower adjacent to water. Therefore, water and air can be distinguished by the sensor. Taped to the outside of a polypropylene beaker, the sensor was able to determine the height of water inside. A glass container with a thickness of 3 mm was also tested successfully. However, response time and sensitivity are in this case reduced due to the higher thermal mass of the glass.

Water and oil have different thermal properties and, therefore, the sensor was tested to distinguish between them. When water and oil are in the same container, the oil floats atop the water, forming two interfaces: water-oil and oil-air. The liquid level sensor was able to determine the height of both interfaces. To further investigate the sensitivity of the liquid level sensor, its ability to sense the segregation of a water-oil mixture was tested. To this end, water and oil were mixed with a mixing agent and poured into a container with the sensor attached on the outer side. With this setup, the signal of the sensor changed depending on the height of its pixels The signals of the pixels increased one by one from the bottom up to roughly the middle, whereas the others' signals decreased. As more water fell out of the mixture, a water-enriched phase formed at the bottom of the container. Water has a higher specific heat capacity than oil, can thus absorb more heat, and yields a higher signal. The oil-enriched phase floating at the top consequently yields lower signals. Thus, the segregation of oil and water could be observed in real time by the external liquid level sensor.

Another possible application for the sensor concept is a surface airflow meter which is introduced in chapter 6. For this application, the sensor is adapted to a 4×4 matrix, providing 2D spatial resolution. Cooling of the heated PTC is not achieved by material with different thermal properties as in the level sensor above, but by different velocities of the same material. The power consumption of a single PTC sample correlates with the airflow around it. As a result, the signal of the sensors pixels vary with airflow. For air speeds below $2\,\mathrm{m/s}$, the sensor exhibited good sensitivity. It was able to detect both airflow perpendicular to and concentrated on one pixel as well as airflow over the surface of the sensor. Response times were in the range of multiple seconds. Applications requiring fast response, i.e. stall warning systems, are thus not suited without further optimization. However, applications with less demand on the time scale are well within the sensors capabilities. It can therefore be assumed that the airflow sensor would perform well for mapping airflow over models in wind tunnels, though further testing is necessary.

In chapter 7, the semiconductor XPRD01B06 was characterized. XPRD01B06 was investigated with XPS to determine its chemical structure. It is composed of five thiophene units per DPP monomer and side chains with 32 ± 4 carbon atoms in total. Batch-to-batch variations were tested over three batches and found to be small. One batch, denoted as 'OP3', exhibited an increased nitrogen content. As the difference is small, impact was expected to be insignificant. OFETs out of this material were also characterized. The contact resistance of these OFETs was determined to be below $20\,\mathrm{k\Omega}$ while channel resistances where in the order of hundreds of kilohm. It was thus concluded that contact resistances have negligible influence on device performance and don't require optimization. The OFETs were also subjected to operational stress. Threshold voltage V_th, on-current I_on, and mobility μ shifted over time under stress. However, either reverse bias or long down time compensated for the drift, returning the parameters to their initial values. An oscillation in the parameters could be seen over the course of a day. This was attributed to fluctuating lab temperature. As the OFETs are - amongst other things - intended for use in an active temperature sensor matrix, the temperature behavior is important and was investigated. In the beginning, both on and off currents depended significantly on temperature. Especially between $40\,^\circ\mathrm{C}$ and $70\,^\circ\mathrm{C}$, off-currents were too strong to be considered 'off'. After process changes by the fabrication team based on these results, the on-off behavior over temperature of the OFETs stabilized and was deemed sufficient for the intended active matrices.

The sensors described in chapters 5 and 6, however, are now at a point of proved concept. With thermal sensing, the liquid level sensor offers an alternative to existing level sensors. Especially for metallic liquid containers which hinder capacitive level sensing, for example, the thermal approach of this work allows for external application. Unfortunately, due to the nature of the thermal approach, a well controlled environment with stable thermal properties is required. To avoid this problem, the sensor could be encapsulated and placed into a liquid container, multiplying sensitivity. Although an external sensor has its advantages, this would allow for detailed and faster resolving of liquids and mixtures. The airflow sensor offers an alternative for smoke or pressure sensitive paint in wind tunnels. Thanks to its flexibility, the airflow sensor can adjust to most surfaces and forms. Dropping the CL, the design is less cluttered. The sensor is capable of mapping surface airflow without placing obstacles in the air's path or requiring optical observation. Although not tested in detail, the sensor matrix also has potential for thermal radiation or light mapping. The spot of a handheld laser pointer could be detected when it hit a pixel of the matrix.

In summary, this work introduces a new sensor concept with promising results as a liquid level and airflow sensor. While successfully demonstrated, some further optimization for commercial use is required. With further testing, more sensing applications, e.g. light mapping, can likely be realized. For joint efforts to develop an active matrix for sensors, OFETs were characterized until they could be fabricated with desired the characteristics.

A. Appendix

A.1. Code

In the following the self-written code used for data collection and visualization are given. They are in no way optimized and may not saturate high code standards but they work as intended and are no bottleneck to the systems used. The readout can reach well above 50 Hz which is much faster than the reaction time of the sensor arrays. Therefore, optimization was forgone. To reduce reading speeds and therefore the amount of and memory required for data, the TIMESTEP in line 2 in the Arduino code can be adjusted.

Arduino Code

```
1  #define    readout_version "v_2.0"
2  #define    TIMESTEP 1
3
4  unsigned long   preMillis = 0;
5
6  void setup() {
7    // initialize serial communication
8    Serial.begin(115200);
9    // set analog pins to input
10   for (int i = A0; i <= A15; i++) {
11     pinMode(i, INPUT);
12   }
13   // wait for serial input
14   pinMode(LED_BUILTIN, OUTPUT);
15   while (!Serial.available()) {
16     digitalWrite(LED_BUILTIN, HIGH);
17     delay(500);
18     digitalWrite(LED_BUILTIN, LOW);
19     delay(500);
20   }
21   // define starting point
22   preMillis = millis();
23 }
24
25 void loop() {
26   // wait for serial input
27   while (!Serial.available()) {
28     digitalWrite(LED_BUILTIN, HIGH);
29     delay(100);
30     digitalWrite(LED_BUILTIN, LOW);
31     delay(100);
32   }
33   String ser_in = Serial.readStringUntil('\n');
```

```
34    // answer or react to input
35    if (ser_in == "version") {Serial.println(readout_version);}
36    // adjust analog reference
37    else if (ser_in == "ref5V") {
38      analogReference(DEFAULT);
39      Serial.println("ref_update");
40    }
41    else if (ser_in == "ref2V56") {
42      analogReference(INTERNAL2V56);
43      Serial.println("ref_update");
44    }
45    else if (ser_in == "ref1V1") {
46      analogReference(INTERNAL1V1);
47      Serial.println("ref_update");
48    }
49    // start sending data
50    else if (ser_in == "go") {
51      digitalWrite(LED_BUILTIN, LOW);
52      sending_data();
53    }
54    // command not understood
55    else {Serial.println("?");}
56  }
57
58  void sending_data() {
59    int keepGoing = 1;
60
61    // send data until aborted
62    while (true) {
63      // exit reading and sending loop
64      if (Serial.available()) {
65        if (Serial.readStringUntil('\n') == "halt") {keepGoing = 0;}
66      }
67      if (keepGoing == 0) {break;}
68
69      // read analog inputs
70      int val0 = analogRead(A0);
71      if (val0 > 1000) {overvoltage_protection(0);}
72      int val1 = analogRead(A1);
73      if (val1 > 1000) {overvoltage_protection(1);}
74      int val2 = analogRead(A2);
75      if (val2 > 1000) {overvoltage_protection(2);}
76      int val3 = analogRead(A3);
77      if (val3 > 1000) {overvoltage_protection(3);}
78      int val4 = analogRead(A4);
79      if (val4 > 1000) {overvoltage_protection(4);}
80      int val5 = analogRead(A5);
81      if (val5 > 1000) {overvoltage_protection(5);}
82      int val6 = analogRead(A6);
83      if (val6 > 1000) {overvoltage_protection(6);}
84      int val7 = analogRead(A7);
85      if (val7 > 1000) {overvoltage_protection(7);}
86      int val8 = analogRead(A8);
87      if (val8 > 1000) {overvoltage_protection(8);}
88      int val9 = analogRead(A9);
```

```
89          (val9 > 1000) {overvoltage_protection(9);}
90          valA = analogRead(A10);
91          (valA > 1000) {overvoltage_protection(10);}
92          valB = analogRead(A11);
93          (valB > 1000) {overvoltage_protection(11);}
94          valC = analogRead(A12);
95          (valC > 1000) {overvoltage_protection(12);}
96          valD = analogRead(A13);
97          (valD > 1000) {overvoltage_protection(13);}
98          valE = analogRead(A14);
99          (valE > 1000) {overvoltage_protection(14);}
100         valF = analogRead(A15);
101         (valF > 1000) {overvoltage_protection(15);}
102
103     // print to serial
104         (millis() - preMillis > TIMESTEP) {
105       preMillis = millis();
106       Serial.print(val0);
107       Serial.print(",");
108       Serial.print(val1);
109       Serial.print(",");
110       Serial.print(val2);
111       Serial.print(",");
112       Serial.print(val3);
113       Serial.print(",");
114       Serial.print(val4);
115       Serial.print(",");
116       Serial.print(val5);
117       Serial.print(",");
118       Serial.print(val6);
119       Serial.print(",");
120       Serial.print(val7);
121       Serial.print(",");
122       Serial.print(val8);
123       Serial.print(",");
124       Serial.print(val9);
125       Serial.print(",");
126       Serial.print(valA);
127       Serial.print(",");
128       Serial.print(valB);
129       Serial.print(",");
130       Serial.print(valC);
131       Serial.print(",");
132       Serial.print(valD);
133       Serial.print(",");
134       Serial.print(valE);
135       Serial.print(",");
136       Serial.print(valF);
137       Serial.print("\n");
138     }
139   }
140   Serial.flush();
141   delay(100);
142   Serial.println("Stopping readout.");
143       0;
```

```
144  }
145
146  void overvoltage_protection(int pin) {
147    // stop reading due to overvoltage
148    while (true) {
149      Serial.println("Overvoltage detected on pin " + String(pin) + ".\
            nLower Voltage and reset.");
150      digitalWrite(LED_BUILTIN, HIGH);
151      delay(1000);
152      digitalWrite(LED_BUILTIN, LOW);
153      delay(1000);
154    }
155  }
```

Visualization Script

`IndiMatrixReadout-v2.0.ino` in line 8 refers to the Arduino script above.

```python
__author__ = 'Rainer B.'

# Rainer's In-Situ Connector
# RISC
# Readout for PTC matrix
# version 4.1.2
#
# to use with IndiMatrixReadout-v2.0.ino

import tkinter as tk
import sys, serial
import time
import cv2
from PIL import Image, ImageTk

version = 'v4.1.2'
readout_version = 'v_2.0'

# interface class
class GUI:
    # constructor
    def __init__(self):
        self.root = tk.Tk()

        # variables
        self.ref_voltage = tk.DoubleVar(value=5)
        self.matrix_type = tk.StringVar(value='matrix')
        self.matrix_type_inv = tk.BooleanVar(value=False)
        self.baseline = tk.BooleanVar(value=False)
        self.conversion_factor = []
        self.baselines = []
        self.instruction = "none"
        self.pixels = []
        self.videoFeed = tk.BooleanVar(value=False)
        self.colorRange = 0.4

        # root window
        self.root.title("RISC " + version)
        self.root.geometry("410x250")
        self.root.resizable(0,0)
        self.root.configure(background='black')
        self.draw_interface()

        # self.root.mainloop()

    # draw interface
    def draw_interface(self):
        # comment in file
        tk.Label(self.root, text="comment for file", fg='#FFFFFF', bg=
            '#000000').place(x=10, y=10)
        self.comment_entry = tk.Entry(self.root)
```

```
51          self.comment_entry.insert(-1, 'no comment')
52          self.comment_entry.place(x=10, y=35, width=100, height=28)
53
54          # connection
55          tk.Label(self.root, text="Arduino port", fg='#FFFFFF', bg='
               #000000').place(x=175, y=10)
56          self.port_entry = tk.Entry(self.root)
57          self.port_entry.insert(tk.END, 'COM7')
58          self.port_entry.place(x=150, y=35, width=50, height=28)
59          self.buttonConnect = tk.Button(self.root, text="connect",
               command=self.connect)
60          self.buttonConnect.place(x=210, y=35, width=60, height=28)
61
62          # reference voltage
63          tk.Label(self.root, text="reference voltage", fg='#FFFFFF', bg
               ='#000000').place(x=300, y=10)
64          self.radButtonRefVolt = []
65          self.radButtonRefVolt.append(tk.Radiobutton(self.root, text="5
                V", variable=self.ref_voltage, indicatoron=False, value=5,
                width=8, command=self.change_reference))
66          self.radButtonRefVolt.append(tk.Radiobutton(self.root, text="2
                V56", variable=self.ref_voltage, indicatoron=False, value
               =2.56, width=8, command=self.change_reference))
67          self.radButtonRefVolt.append(tk.Radiobutton(self.root, text="1
                V1", variable=self.ref_voltage, indicatoron=False, value
               =1.1, width=8, command=self.change_reference))
68          for button in self.radButtonRefVolt:
69              button.place(x=350, y=35+25*self.radButtonRefVolt.index(
                   button), width=50, height=28)
70              button.config(state='disabled')
71
72          # matrix layout
73          tk.Label(self.root, text="pixel layout", fg='#FFFFFF', bg='
               #000000').place(x=10, y=85)
74          self.radButtonLayout = []
75          self.radButtonLayout.append(tk.Radiobutton(self.root, text='
               matrix', variable=self.matrix_type, indicatoron=False,
               value="matrix", width=8))
76          self.radButtonLayout.append(tk.Radiobutton(self.root, text="
               level", variable=self.matrix_type, indicatoron=False, value
               ="level", width=8))
77          self.checkboxLayoutInverse = tk.Checkbutton(self.root, text="
               inv", variable=self.matrix_type_inv, fg='#FFFFFF', bg='
               #000000', selectcolor='#000000', activeforeground='#FFFFFF'
               , activebackground='#000000')
78          for button in self.radButtonLayout:
79              button.place(x=10, y=110+25*self.radButtonLayout.index(
                   button), width=50, height=28)
80          self.checkboxLayoutInverse.place(x=60, y=135, width=40, height
               =28)
81
82
83          # start and stop readout
84          self.buttonStart = tk.Button(self.root, text="start", bg='#00
               FF00', command=self.start_recording, state='disabled')
```

```python
85         self.buttonStart.place(x=150, y=120, width=50, height=28)
86         self.buttonStop = tk.Button(self.root, text="stop", command=
               self.stop_recording, state='disabled')
87         self.buttonStop.place(x=210, y=120, width=50, height=28)
88
89         # record screen
90         tk.Button(self.root, text="record", state='disabled').place(x
               =350, y=135, width=50, height=28)
91
92         # video feed
93         tk.Label(self.root, text="enable video", fg='#FFFFFF', bg='
               #000000').place(x=10, y=185)
94         self.radButtonVideo = []
95         self.radButtonVideo.append(tk.Radiobutton(self.root, text='on'
               , variable=self.videoFeed, indicatoron=False, value=True,
               width=5))
96         self.radButtonVideo.append(tk.Radiobutton(self.root, text='off
               ', variable=self.videoFeed, indicatoron=False, value=False,
               width=5))
97         for button in self.radButtonVideo:
98             button.place(x=10+30*self.radButtonVideo.index(button), y
                   =210, width=30, height=28)
99             button.config(state='disabled')
100
101        # baseline
102        self.buttonBaseline = tk.Button(self.root, text="record
               baseline", command=self.get_baseline, state='disabled')
103        self.buttonBaseline.place(x=155, y=185, width=100, height=28)
104        self.radButtonBaseline = []
105        self.radButtonBaseline.append(tk.Radiobutton(self.root, text="
               on", variable=self.baseline, indicatoron=False, value=True,
               width=8))
106        self.radButtonBaseline.append(tk.Radiobutton(self.root, text="
               off", variable=self.baseline, indicatoron=False, value=
               False, width=8))
107        for button in self.radButtonBaseline:
108            button.place(x=155+50*self.radButtonBaseline.index(button)
                   , y=210, width=50, height=28)
109            button.config(state='disabled')
110
111        # reset applicaton
112        tk.Button(self.root, text="reset", command=self.reset).place(x
               =350, y=185, width=50, height=28)
113
114        # close application
115        tk.Button(self.root, text="close", command=self.close).place(x
               =350, y=210, width=50, height=28)
116
117    # draw pixels as matrix assembly
118    def draw_matrix(self, pixelsX = 4, pixelsY = 4, pixelSize=100,
           border=20):
119        geometryString = ''
120        if self.videoFeed.get():
121            geometryString = str(pixelsX*pixelSize+pixelsY*pixelSize
                   +3*border) + 'x' + str(pixelsY*pixelSize+2*border)
```

```python
122         else:
123             geometryString =    str(pixelsX*pixelSize+2*border) + 'x' +
                    str(pixelsY*pixelSize+2*border)
124         self.matrix = tk.Toplevel()
125         self.matrix.title('Matrix')
126         self.matrix.geometry(geometryString + '+50+50')
127         self.matrix.resizable(0,0)
128         self.matrix.configure(background='black')
129         self.pixels = [tk.Label(master=self.matrix, bg='#0000FF', fg='
                #FFFFFF') for x in range(pixelsX * pixelsY)]
130         for pixel in self.pixels:
131             pixel.place(x=border+(self.pixels.index(pixel)%pixelsX)*
                    pixelSize, y=border+(self.pixels.index(pixel)//pixelsX)
                    *pixelSize, width=pixelSize, height=pixelSize)
132         if self.videoFeed.get():
133             self.videoImg = tk.Label(self.matrix, bg='#000000')
134             self.videoImg.place(x=pixelsX*pixelSize+2*border, y=border
                    , width=pixelsY*pixelSize, height=pixelsY*pixelSize)

136     # draw pixels as one line
137     def draw_level(self, pixelsX = 16, pixelSize=35, border=20):
138         geometryString = ''
139         if self.videoFeed.get():
140             geometryString =    str(2*pixelSize+pixelsX*pixelSize+3*
                    border) + 'x' +    str(pixelsX*pixelSize+2*border)
141         else:
142             geometryString =    str(2*pixelSize+2*border) + 'x' +    str(
                    pixelsX*pixelSize+2*border)
143         self.level = tk.Toplevel()
144         self.level.title('Level')
145         self.level.geometry(geometryString + '+50+50')
146         self.level.resizable(0,0)
147         self.level.configure(background='black')
148         self.pixels = [tk.Label(master=self.level, bg='#0000FF', fg='#
                FFFFFF') for x in range(pixelsX)]
149         for pixel in self.pixels:
150             if not self.matrix_type_inv.get():
151                 pixel.place(x=border, y=border+(pixelsX-1-self.pixels.
                        index(pixel))*pixelSize, width=pixelSize*2, height=
                        pixelSize)
152             else:
153                 pixel.place(x=border, y=border+(self.pixels.index(
                        pixel))*pixelSize, width=pixelSize*2, height=
                        pixelSize)
154         if self.videoFeed.get():
155             self.videoImg = tk.Label(self.level, bg='#000000')
156             self.videoImg.place(x=2*pixelSize+2*border, y=border,
                    width=pixelsX*pixelSize, height=pixelsX*pixelSize)

158     # update pixels according to data
159     def update_drawing(self, data):
160         for x in range(len(self.pixels)):
161             value = data[x]
162             self.pixels[x].configure(bg=self.intToColor(number = value
                    , id = x), text="{raw}\n{signal:.2f}".format(raw=value,
```

```python
                         signal=value*self.conversion_factor[x]))
        self.root.update()
        return self.running

    # convert integer to color string
    def intToColor(self, number, id):
        self.colorString = "#"
        if (self.baseline.get() == False):
            if (number<0 or number>1023):
                self.colorString = "#FFFF00"
            elif (number>511):
                self.colorString = "#0000{:02x}".format(int((number
                    -512)/2))  # 0 ... 255
            else:
                self.colorString = "#{:02x}0000".format(int((511-
                    number)/2))  # 0 ... 255
        else:
            relative = number * self.conversion_factor[id]
            if (relative<(1-self.colorRange)):
                self.colorString = "#FFFF00"
            elif (relative>(1+self.colorRange)):
                self.colorString = "#0000FF"
            else:
                self.colorString = "#{:02x}{:02x}{:02x}".format(int
                    (128-(relative-(1-self.colorRange))*128/(2*self.
                    colorRange+0.01)), int(128-(relative-(1-self.
                    colorRange))*128/(2*self.colorRange+0.01)), int
                    (128+(relative-(1-self.colorRange))*128/(2*self.
                    colorRange+0.01)))
        return self.colorString

    # run tkinter mainloop to set up measurement
    def wait_for_input(self):
        print("Waiting for user input.")
        self.root.mainloop()
        return self.instruction

    # connect to readout
    def connect(self, status = None):
        if (status == None):
            self.instruction = "connect"
            self.root.quit()
            return 0
        elif (status == 0):
            self.port_entry.config(disabledbackground='#00FF00', state
                ='disabled')
            self.buttonConnect.config(state='disabled')
        elif (status == 1):
            self.port_entry.config(disabledbackground='#FFFF00', state
                ='disabled')
            self.buttonConnect.config(state='disabled')
        elif (status == -1):
            self.port_entry.delete(0, tk.END)
            self.port_entry.insert(0, "Error")
            return -1
```

```python
208            for thing in self.radButtonRefVolt + [self.buttonBaseline,
                   self.buttonStart] + self.radButtonVideo:
209                thing.config(state='normal')
210
211        # change reference voltage
212        def change_reference(self):
213            self.instruction = "reference"
214            self.baseline.set(False)
215            self.baselines = []
216            for button in self.radButtonBaseline:
217                button.config(state='disabled')
218            self.root.quit()
219
220        # get baseline data
221        def get_baseline(self, data = None):
222            # initiate baseline gathering
223            if (data == None):
224                self.buttonBaseline.config(bg='#FFFF00')
225                self.root.update()
226                self.instruction = "baseline"
227                self.root.quit()
228            # set baseline
229            else:
230                self.buttonBaseline.config(bg='SystemButtonFace')
231                if (0.0 in data):
232                    print("Baseline contains zero(s). Skipping basline.")
233                    return 1
234                self.baselines = data
235                self.baseline.set(True)
236                print("Baselines set.")
237                for button in self.radButtonBaseline:
238                    button.config(state='normal')
239
240        # start measurement
241        def start_recording(self):
242            for thing in [self.comment_entry] + self.radButtonRefVolt +
                   self.radButtonLayout + [self.buttonBaseline, self.
                   buttonStart] + self.radButtonBaseline:
243                thing.config(state='disabled')
244            self.buttonStop.config(state='normal')
245            self.running = 1
246            if (self.matrix_type.get() == "matrix"):
247                self.draw_matrix()
248            elif (self.matrix_type.get() == "level"):
249                self.draw_level()
250            if (self.baseline.get() == True):
251                self.conversion_factor = [1/x for x in self.baselines]
252            else:
253                self.conversion_factor = [self.ref_voltage.get()/1024 for
                   x in range(len(self.pixels))]
254            self.instruction = "measurement"
255            self.root.quit()
256
257        # stop recording
258        def stop_recording(self, change = None):
```

```python
259                 self.running = 0
260             if (change != None):
261                 for thing in [self.comment_entry] + self.radButtonRefVolt
                        + self.radButtonLayout + [self.buttonBaseline, self.
                        buttonStart]:
262                     thing.config(state='normal')
263                 if self.baselines:
264                     for button in self.radButtonBaseline:
265                         button.config(state='normal')
266             self.buttonStop.config(state='disabled')
267             try:
268                 self.matrix.destroy()
269             except:
270                 pass
271             try:
272                 self.level.destroy()
273             except:
274                 pass
275
276     # reset application
277     def reset(self):
278         self.instruction = "reset"
279         self.root.quit()
280
281     # send command to close
282     def close(self):
283         self.instruction = "close"
284         self.root.quit()
285
286     # clean up
287     def cleanup(self):
288         self.running = 0
289         self.root.quit()
290         self.root.destroy()
291
292 # serial readout class
293 class readout:
294     # constructor
295     def __init__(self, strPort = "COM7", resorting = False,
            voltage_range = 5):
296         # set parameters
297         self.port = strPort
298         self.resorting = resorting
299         self.resortlist = { # account for pin layout of sample
300             0: 9,
301             1: 8,
302             2: 7,
303             3: 6,
304             4: 11,
305             5: 10,
306             6: 5,
307             7: 4,
308             8: 13,
309             9: 12,
310             10: 3,
```

```python
            11: 2,
            12: 15,
            13: 14,
            14: 1,
            15: 0}
        self.voltage_range = voltage_range

    # establish connection to readout
    def connecting(self, baudRate = 115200, maxWaitTime = 5):
        try:
            self.ser = serial.Serial(self.port, baudRate, timeout=
                maxWaitTime)
            print("Connecting to serial port %s..." % self.port)
        except serial.serialutil.SerialException:
            print("Connection error. Check connection.")
            return -1
        except:
            print("Error 01")
            return -1
        time.sleep(1) # time to establish connection

        # check for correct version
        self.ser.reset_input_buffer()
        print("--> version")
        self.ser.write(b'version\n')
        # time.sleep(1)
        answer = self.ser.readline().decode('utf-8')
        print("<-- " + answer, end='')
        if (answer != readout_version + "\r\n"):
            print("Wrong readout version.")
            return 1
        else:
            return 0

    # set reference voltage
    def set_reference(self, voltage_range = None):
        if voltage_range != None:
            self.voltage_range = voltage_range
        if (self.voltage_range == 5):
            print("--> ref5V")
            self.ser.write(b'ref5V\n')
        elif (self.voltage_range == 2.56):
            print("--> ref2V56")
            self.ser.write(b'ref2V56\n')
        elif (self.voltage_range == 1.1):
            print("--> ref1V1")
            self.ser.write(b'ref1V1\n')
        answer = self.ser.readline().decode('utf-8')
        print("<-- " + answer, end='')
        if (answer == 'ref_update' + "\r\n"):
            return 0
        else:
            return 1

    # get baseline reading
```

```
365        def get_baseline(self, period = 5):
366            try:
367                self.start_readout()
368                data = self.read_data()
369                number_of_points = 1
370                while (time.perf_counter() < self.start_time + period):
371                    data_new = self.read_data()
372                    for i in range(len(data)):
373                        data[i] += data_new[i]
374                    number_of_points += 1
375                self.pause_read()
376                for i in range(len(data)):
377                    data[i] = data[i] / number_of_points
378                return data
379            except:
380                print("Error. Couldn't get baseline.")
381                return -1
382
383        # request data flow
384        def start_readout(self):
385            print("Starting readout...")
386            print("--> go")
387            self.ser.write(b'go\n')
388            # request data and empty buffer
389            # self.ser.reset_input_buffer()
390            self.start_time = time.perf_counter()
391
392        # halt data flow
393        def pause_read(self):
394            print("--> pause")
395            self.ser.write(b'halt\n')
396            time.sleep(0.05)
397            self.ser.reset_input_buffer()
398            print("<-- " + self.ser.readline().decode('utf-8'), end='')
399
400        # read and return data
401        def read_data(self, file = None, matrix_type = None):
402            if (file != None):
403                file.write("{},".format(time.perf_counter() - self.
                    start_time))
404            try:
405                self.ser.reset_input_buffer()
406                self.ser.readline()
407                line = self.ser.readline()
408                if (line.startswith("Overvoltage".encode('utf-8'))):
409                    if (file != None):
410                        print(line.decode('utf-8'))
411                    print(line.decode('utf-8'), end='')
412                    return 1
413                data = [int(val) for val in line.split(b',')]
414                if (self.resorting == True):
415                    data2 = []
416                    for i in range(16):
417                        data2.append(data[self.resortlist.get(i)])
418                    if (file != None):
```

```python
419                             file.write(",".join(str(item) for item in data2) +
                                    '\n')
420                         return data2
421                 if (file != None):
422                     file.write(line.decode('utf-8'))
423                 return data
424             except ValueError:
425                 print("Communication error.")
426                 return -1
427             except KeyboardInterrupt:
428                 print("User interrupt.")
429                 return -1

431     # clean up
432     def close(self):
433         print("Closing connection to %s..." % self.port)
434         self.ser.flush()
435         self.ser.close()

437 # backend
438 def main():
439     interface = GUI()
440     matrix = readout()
441     while True:
442         instruction = interface.wait_for_input()
443         print("User input: " + instruction)
444         if (instruction == 'connect'):
445             matrix.port = interface.port_entry.get()
446             status = matrix.connecting()
447             interface.connect(status = status)
448         elif (instruction == 'reference'):
449             matrix.set_reference(interface.ref_voltage.get())
450         elif (instruction == 'baseline'):
451             baseline = matrix.get_baseline()
452             interface.get_baseline(baseline)
453         elif (instruction == 'measurement'):
454             if (interface.matrix_type.get() == 'matrix'):
455                 matrix.resorting = True
456             else:
457                 matrix.resorting = False
458             if (interface.videoFeed.get()):
459                 cap = cv2.VideoCapture(0, cv2.CAP_DSHOW)
460             matrix.start_readout()
461             ofile = open("matrixReadout_{}_{}.dat".format(time.
                    strftime("%Y%m%d-%H%M%S"), interface.comment_entry.get
                    ()), 'x')
462             ofile.write("### reference voltage: {}, pixel arrangement:
                    {}\n".format(str(matrix.voltage_range), interface.
                    matrix_type.get()))
463             try:
464                 # keep the loop running
465                 while True:
466                     newData = matrix.read_data(ofile)
467                     if (newData == -1 or newData == 1):
468                         break
```

```python
            if (interface.update_drawing(newData) == 0):
                break
            if (interface.videoFeed.get()):
                # read video feed of webcam
                image = cap.read()[1]
                # do conversions of the image for tkinter
                image = cv2.cvtColor(image, cv2.COLOR_BGR2RGB)
                image = Image.fromarray(image)
                image = ImageTk.PhotoImage(image)
                # finally display the image
                interface.videoImg.configure(image = image)
        except KeyboardInterrupt:
            pass

        # clean file saving
        try:
            ofile.close()
        except:
            print("Error saving file.")
        interface.stop_recording(change = True)
        matrix.pause_read()
        if (interface.videoFeed.get()):
            cap.release() # release hardware
            cv2.destroyAllWindows() # necessary?
    elif (instruction == "reset"):
        try:
            interface.cleanup()
            matrix.close()
        except:
            pass
        del interface, matrix
        time.sleep(0.5)
        interface = GUI()
        matrix = readout()
    elif (instruction == "close"):
        break
try:
    interface.cleanup()
    matrix.close()
except:
    pass

# call main
if __name__ == '__main__':
    main()
```

Bibliography

[1] Henry Alexander Ignatious, Hesham-El- Sayed, and Manzoor Khan. 'An overview of sensors in Autonomous Vehicles'. In: *Procedia Computer Science* 198.2021 (2022), pp. 736–741. DOI: 10.1016/j.procs.2021.12.315.

[2] Jorge Vargas, Suleiman Alsweiss, Onur Toker, Rahul Razdan, and Joshua Santos. 'An Overview of Autonomous Vehicles Sensors and Their Vulnerability to Weather Conditions'. In: *Sensors* 21.16 (Aug. 2021), p. 5397. DOI: 10.3390/s21165397.

[3] Garmin Ltd. *Garmin fēnix® 7 – Sapphire Solar Edition | Multisport GPS Smartwatch.* URL: https://www.garmin.com/en-GB/p/735520/#specs (visited on 04/27/2023).

[4] Jane McCann and David Bryson, eds. *Smart Clothes and Wearable Technology.* 2nd. Sawston, Cambridge, UK: Woodhead Publishing, 2022. ISBN: 9780128205778.

[5] Sofia Scataglini, A.P. Moorhead, and F. Feletti. 'A Systematic Review of Smart Clothing in Sports: possible Applications to Extreme Sports'. In: *Muscle Ligaments and Tendons Journal* 10.02 (June 2020), p. 333. DOI: 10.32098/mltj.02.2020.19.

[6] K. Kudo and T. Moriizumi. 'Spectrum-controllable color sensors using organic dyes'. In: *Applied Physics Letters* 39.8 (Oct. 1981), pp. 609–611. DOI: 10.1063/1.92820.

[7] Jan J. Miasik, Alan Hooper, and Bruce C. Tofield. 'Conducting polymer gas sensors'. In: *Journal of the Chemical Society, Faraday Transactions 1: Physical Chemistry in Condensed Phases* 82.4 (1986), p. 1117. DOI: 10.1039/f19868201117.

[8] William R. Heineman, Henry J. Wieck, and Alexander M. Yacynych. 'Polymer film chemically modified electrode as a potentiometric sensor'. In: *Analytical Chemistry* 52.2 (Feb. 1980), pp. 345–346. DOI: 10.1021/ac50052a031.

[9] P. Leaver, M.J. Cunningham, and B.E. Jones. 'Piezoelectric polymer pressure sensors'. In: *Sensors and Actuators* 12.3 (Oct. 1987), pp. 225–233. DOI: 10.1016/0250-6874(87)80036-7.

[10] *The Nobel Prize in Chemistry 2000.* 2000. URL: https://www.nobelprize.org/prizes/chemistry/2000/summary/ (visited on 04/18/2023).

[11] J. Spadavecchia, G. Ciccarella, P. Siciliano, S. Capone, and R. Rella. 'Spin-coated thin films of metal porphyrin–phthalocyanine blend for an optochemical sensor of alcohol vapours'. In: *Sensors and Actuators B: Chemical* 100.1-2 (June 2004), pp. 88–93. DOI: 10.1016/j.snb.2003.12.027.

[12] Donggeon Han, Yasser Khan, Jonathan Ting, Simon M. King, Nir Yaacobi-Gross, Martin J. Humphries, Christopher J. Newsome, and Ana C. Arias. 'Flexible Blade-Coated Multicolor Polymer Light-Emitting Diodes for Optoelectronic Sensors'. In: *Advanced Materials* 29.22 (June 2017), p. 1606206. DOI: 10.1002/adma.201606206.

[13] Giorgio Mattana and Danick Briand. 'Recent advances in printed sensors on foil'. In: *Materials Today* 19.2 (Mar. 2016), pp. 88–99. DOI: 10.1016/j.mattod.2015.08.001.

[14] Ana Moya, Gemma Gabriel, Rosa Villa, and F. Javier del Campo. 'Inkjet-printed electrochemical sensors'. In: *Current Opinion in Electrochemistry* 3.1 (June 2017), pp. 29–39. DOI: 10.1016/j.coelec.2017.05.003.

[15] Jie Dai, Osarenkhoe Ogbeide, Nasiruddin Macadam, Qian Sun, Wenbei Yu, Yu Li, Bao-Lian Su, Tawfique Hasan, Xiao Huang, and Wei Huang. 'Printed gas sensors'. In: *Chemical Society Reviews* 49.6 (2020), pp. 1756–1789. DOI: 10.1039/C9CS00459A.

[16] Jueng Gil (James) Lee, Zhuo Gao, Dong Fu, and Xiaolin Yan. '31" Flexible Printed OLED TV Display Technology: It's TV Mobiles'. In: *SID Symposium Digest of Technical Papers* 53.1 (June 2022), pp. 993–997. DOI: 10.1002/sdtp.15664.

[17] D&O LG Communication Center. *LAUNCH OF LG'S LONG-AWAITED ROLLABLE OLED TV MARKS TURNING POINT IN TV HISTORY.* 2020. URL: https://www.lgcorp.com/media/release/22562 (visited on 04/27/2023).

[18] Joost Kimpel and Tsuyoshi Michinobu. 'Conjugated polymers for functional applications: lifetime and performance of polymeric organic semiconductors in organic field-effect transistors'. In: *Polymer International* 70.4 (Apr. 2021), pp. 367–373. DOI: 10.1002/pi.6020.

[19] Shuguang Wang, Zhongwu Wang, Jie Li, Liqiang Li, and Wenping Hu. 'Surface-grafting polymers: from chemistry to organic electronics'. In: *Materials Chemistry Frontiers* 4.3 (2020), pp. 692–714. DOI: 10.1039/C9QM00450E.

[20] Sebastian Scholz, Denis Kondakov, Björn Lüssem, and Karl Leo. 'Degradation Mechanisms and Reactions in Organic Light-Emitting Devices'. In: *Chemical Reviews* 115.16 (Aug. 2015), pp. 8449–8503. DOI: 10.1021/cr400704v.

[21] C. K. Chiang, C. R. Fincher, Y. W. Park, A. J. Heeger, H. Shirakawa, E. J. Louis, S. C. Gau, and Alan G. MacDiarmid. 'Electrical Conductivity in Doped Polyacetylene'. In: *Physical Review Letters* 39.17 (Oct. 1977), pp. 1098–1101. DOI: 10.1103/PhysRevLett.39.1098.

[22] Chengliang Wang, Huanli Dong, Wenping Hu, Yunqi Liu, and Daoben Zhu. 'Semiconducting π-Conjugated Systems in Field-Effect Transistors: A Material Odyssey of Organic Electronics'. In: *Chemical Reviews* 112.4 (Apr. 2012), pp. 2208–2267. DOI: 10.1021/cr100380z.

[23] W. E. Brittin. 'Valence angle of the tetrahedral carbon atom'. In: *Journal of Chemical Education* 22.3 (Mar. 1945), p. 145. DOI: 10.1021/ed022p145.

[24] J.-C. Charlier and J.-P. Issi. 'Electrical conductivity of novel forms of carbon'. In: *Journal of Physics and Chemistry of Solids* 57.6-8 (June 1996), pp. 957–965. DOI: 10.1016/0022-3697(95)00382-7.

[25] Johannes Thiele. 'Zur Kenntniss der ungesättigten Verbindungen. Theorie der ungesättigten und aromatischen Verbindungen'. In: *Justus Liebig's Annalen der Chemie* 306.1-2 (1899), pp. 87–142. DOI: 10.1002/jlac.18993060107.

[26] Vladsinger. *Benzene Representations*. 2009. URL: https://commons.wikimedia.org/wiki/File:Benzene_Representations.svg (visited on 03/22/2023).

[27] Austin M. Kay, Oskar J. Sandberg, Nasim Zarrabi, Wei Li, Stefan Zeiske, Christina Kaiser, Paul Meredith, and Ardalan Armin. 'Quantifying the Excitonic Static Disorder in Organic Semiconductors'. In: *Advanced Functional Materials* 32.32 (Aug. 2022), p. 2113181. DOI: 10.1002/adfm.202113181.

[28] H. Bässler. 'Charge Transport in Disordered Organic Photoconductors a Monte Carlo Simulation Study'. In: *physica status solidi (b)* 175.1 (Jan. 1993), pp. 15–56. DOI: 10.1002/pssb.2221750102.

[29] Charles Kittel. *Introduction to Solid State Physics*. Ed. by Stuart Johnson. 8th ed. John Wiley and Sons Inc, 2005. ISBN: 0-471-41526-X.

[30] Anna B. Chwang and C. Daniel Frisbie. 'Temperature and gate voltage dependent transport across a single organic semiconductor grain boundary'. In: *Journal of Applied Physics* 90.3 (Aug. 2001), pp. 1342–1349. DOI: 10.1063/1.1376404.

[31] D. Kumar and R.C. Sharma. 'Advances in conductive polymers'. In: *European Polymer Journal* 34.8 (Aug. 1998), pp. 1053–1060. DOI: 10.1016/S0014-3057(97)00204-8.

[32] Pradip Kar. *Doping in Conjugated Polymers*. Wiley, 2013, p. 176. ISBN: 9781118573808.

[33] Karsten Walzer, B. Maennig, Martin Pfeiffer, and Karl Leo. 'Highly Efficient Organic Devices Based on Electrically Doped Transport Layers'. In: *Chemical Reviews* 107.4 (Apr. 2007), pp. 1233–1271. DOI: 10.1021/cr050156n.

[34] Mohamed S. A. Abdou, Francesco P. Orfino, Yongkeun Son, and Steven Holdcroft. 'Interaction of Oxygen with Conjugated Polymers: Charge Transfer Complex Formation with Poly(3-alkylthiophenes)'. In: *Journal of the American Chemical Society* 119.19 (May 1997), pp. 4518–4524. DOI: 10.1021/ja964229j.

[35] Marc-Michael Barf, Frank S. Benneckendorf, Patrick Reiser, Rainer Bäuerle, Wolfgang Köntges, Lars Müller, Martin Pfannmöller, Sebastian Beck, Eric Mankel, Jan Freudenberg, Daniel Jänsch, Jean-Nicolas Tisserant, Robert Lovrincic, Rasmus R. Schröder, Uwe H. F. Bunz, Annemarie Pucci, Wolfram Jaegermann, Wolfgang Kowalsky, and Klaus Müllen. 'Compensation of Oxygen Doping in p-Type Organic Field-Effect Transistors Utilizing Immobilized n-Dopants'. In: *Advanced Materials Technologies* 2000556 (Nov. 2020), p. 2000556. DOI: 10.1002/admt.202000556.

[36] Lewis M. Cowen, Jonathan Atoyo, Matthew J. Carnie, Derya Baran, and Bob C. Schroeder. 'Review—Organic Materials for Thermoelectric Energy Generation'. In: *ECS Journal of Solid State Science and Technology* 6.3 (2017), N3080–N3088. DOI: 10.1149/2.0121703jss.

[37] R. D. Sherman, L. M. Middleman, and S. M. Jacobs. 'Electron transport processes in conductor-filled polymers'. In: *Polymer Engineering and Science* 23.1 (Jan. 1983), pp. 36–46. DOI: 10.1002/pen.760230109.

[38] Mostafizur Rahaman, Tapan Kumar Chaki, and Dipak Khastgir. 'Control of the temperature coefficient of the DC resistivity in polymer-based composites'. In: *Journal of Materials Science* 48.21 (Nov. 2013), pp. 7466–7475. DOI: 10.1007/s10853-013-7561-9.

[39] Sadie I. White, Rose M. Mutiso, Patrick M. Vora, David Jahnke, Sam Hsu, James M. Kikkawa, Ju Li, John E. Fischer, and Karen I. Winey. 'Electrical percolation behavior in silver nanowire-polystyrene composites: Simulation and experiment'. In: *Advanced Functional Materials* 20.16 (2010), pp. 2709–2716. DOI: 10.1002/adfm.201000451.

[40] Keizo Miyasaka, Kiyosi Watanabe, Eiichiro Jojima, Hiromi Aida, Masao Sumita, and Kinzo Ishikawa. 'Electrical conductivity of carbon-polymer composites as a function of carbon content'. In: *Journal of Materials Science* 17.6 (1982), pp. 1610–1616. DOI: 10.1007/BF00540785.

[41] Yao Huang, Semen Kormakov, Xiaoxiang He, Xiaolong Gao, Xiuting Zheng, Ying Liu, Jingyao Sun, and Daming Wu. 'Conductive Polymer Composites from Renewable Resources: An Overview of Preparation, Properties, and Applications'. In: *Polymers* 11.2 (Jan. 2019), p. 187. DOI: 10.3390/polym11020187.

[42] Julian Nagel, Thomas Hanemann, Bastian E Rapp, and Guido Finnah. 'Enhanced PTC Effect in Polyamide/Carbon Black Composites'. In: *Materials* 15.15 (Aug. 2022), p. 5400. DOI: 10.3390/ma15155400.

[43] Goce L. Arsov. 'Celebrating 65th Anniversary of the Transistor'. In: *Electronics ETF* 17.2 (Dec. 2013), pp. 63–70. DOI: 10.7251/ELS1317063A.

[44] Muhammed Conger and İbrahim Develi. 'Innovative 3D Heterogeneous Chip Manufacturing Approach to the Problem of Approaching Physical Limits with Traditional Chip Manufacturing Technologies'. In: *International Conference on Recent Academic Studies* 1 (May 2023), pp. 172–178. DOI: 10.59287/icras.691.

[45] Hanyu Jia and Ting Lei. 'Emerging research directions for n-type conjugated polymers'. In: *Journal of Materials Chemistry C* 7.41 (2019), pp. 12809–12821. DOI: 10.1039/C9TC02632K.

[46] Ruisha Hao, Lei Liu, Jiangyan Yuan, Lingli Wu, and Shengbin Lei. 'Recent Advances in Field Effect Transistor Biosensors: Designing Strategies and Applications for Sensitive Assay'. In: *Biosensors* 13.4 (Mar. 2023), p. 426. DOI: 10.3390/bios13040426.

[47] Fábio Vidor, Thorsten Meyers, and Ulrich Hilleringmann. 'Flexible Electronics: Integration Processes for Organic and Inorganic Semiconductor-Based Thin-Film Transistors'. In: *Electronics* 4.3 (July 2015), pp. 480–506. DOI: 10.3390/electronics4030480.

[48] Charles A. Kraus. 'The Temperature Coefficient of Resistance of Metals at Constant Volume and its Bearing on the Theory of Metallic Conduction'. In: *Physical Review* 4.3 (Sept. 1914), pp. 159–162. DOI: 10.1103/PhysRev.4.159.

[49] J. Meyer. 'Stability of polymer composites as positive-temperature-coefficient resistors'. In: *Polymer Engineering and Science* 14.10 (Oct. 1974), pp. 706–716. DOI: 10.1002/pen.760141009.

[50] Jung-Il Kim, Phil Hyun Kang, and Young Chang Nho. 'Positive temperature coefficient behavior of polymer composites having a high melting temperature'. In: *Journal of Applied Polymer Science* 92.1 (Apr. 2004), pp. 394–401. DOI: 10.1002/app.20064.

[51] Wen-long Cheng, Shuai Yuan, and Jia-liang Song. 'Studies on preparation and adaptive thermal control performance of novel PTC (positive temperature coefficient) materials with controllable Curie temperatures'. In: *Energy* 74.C (Sept. 2014), pp. 447–454. DOI: 10.1016/j.energy.2014.07.009.

[52] Jin Jeon, Han-Bo-Ram Lee, and Zhenan Bao. 'Flexible Wireless Temperature Sensors Based on Ni Microparticle-Filled Binary Polymer Composites'. In: *Advanced Materials* 25.6 (Feb. 2013), pp. 850–855. DOI: 10.1002/adma.201204082.

Bibliography

[53] Yi Liu, Eric Asare, Harshit Porwal, Ettore Barbieri, Stergios Goutianos, Jamie Evans, Mark Newton, James J.C. Busfield, Ton Peijs, Han Zhang, and Emiliano Bilotti. 'The effect of conductive network on positive temperature coefficient behaviour in conductive polymer composites'. In: *Composites Part A: Applied Science and Manufacturing* 139.June (Dec. 2020), p. 106074. DOI: 10.1016/j.compositesa.2020.106074.

[54] G. F. Strouse. *Standard Platinum Resistance Thermometer Calibrations from the Ar TP to the Ag FP*. Tech. rep. 81. Gaithersburg, MD: National Institute of Standards and Technology, Jan. 2008. DOI: 10.6028/NIST.SP.250-81.

[55] Feng Xue, Kangcai Li, Lei Cai, and Enyong Ding. 'Effects of POE and Carbon Black on the PTC Performance and Flexibility of High-Density Polyethylene Composites'. In: *Advances in Polymer Technology* 2021 (Nov. 2021). Ed. by Szczepan Zapotoczny, pp. 1–11. DOI: 10.1155/2021/1124981.

[56] Kazuyuki Ohe and Yoshihide Naito. 'A New Resistor Having An Anomalously Large Positive Temperature Coefficient'. In: *Japanese Journal of Applied Physics* 10.1 (Jan. 1971), p. 99. DOI: 10.1143/JJAP.10.99.

[57] Yi Liu, Han Zhang, Harshit Porwal, Wei Tu, Kening Wan, Jamie Evans, Mark Newton, J. J.C. Busfield, Ton Peijs, and Emiliano Bilotti. 'Tailored pyroresistive performance and flexibility by introducing a secondary thermoplastic elastomeric phase into graphene nanoplatelet (GNP) filled polymer composites for self-regulating heating devices'. In: *Journal of Materials Chemistry C* 6.11 (2018), pp. 2760–2768. DOI: 10.1039/c7tc05621d.

[58] Zhouyao Yue, Huashan Wang, Xiurong Hou, Yuting Shi, and Xin Zhang. 'Enhanced reproducibility of positive temperature coefficient effect of TPO/HDPE blends via elastic crosslinking'. In: *Materials Today Communications* 34.September 2022 (Mar. 2023), p. 105078. DOI: 10.1016/j.mtcomm.2022.105078.

[59] Can Zhou, Yangyang Zhang, Fangjie Cen, Xie Yu, Wenjing Zhou, Shenglin Jiang, and Yan Yu. 'Polymer positive temperature coefficient composites with room-temperature Curie point and superior flexibility for self-regulating heating devices'. In: *Polymer* 265.November 2022 (Jan. 2023), p. 125587. DOI: 10.1016/j.polymer.2022.125587.

[60] Andrzej Rybak, Gisele Boiteux, Flavien Melis, and Gerard Seytre. 'Conductive polymer composites based on metallic nanofiller as smart materials for current limiting devices'. In: *Composites Science and Technology* 70.2 (Feb. 2010), pp. 410–416. DOI: 10.1016/j.compscitech.2009.11.019.

[61] R. Strumpler, J. Skindhoj, J. Glatz-Reichenbach, J.H.W. Kuhlefelt, and F. Perdoncin. 'Novel medium voltage fault current limiter based on polymer PTC resistors'. In: *IEEE Transactions on Power Delivery* 14.2 (Apr. 1999), pp. 425–430. DOI: 10.1109/61.754084.

[62] Henkel. *LOCTITE ECI 8001 E&C*. Düsseldorf, 2017.

[63] David R. Lide, ed. *CRC Handbook of Chemistry and Physics*. 84th. Boca Raton, FL, USA: CRC Press, 2003.

[64] James E. Mark, ed. *Polymer Data Handbook*. Oxford, UK: Oxford University Press, 1999.

[65] Antonella Patti and Domenico Acierno. *Polypropylene - Polymerization and Characterization of Mechanical and Thermal Properties*. Ed. by Weiyu Wang and Yiming Zeng. IntechOpen, May 2020. ISBN: 978-1-83880-414-5. DOI: 10.5772/intechopen.73995.

[66] Massoud Kaviany. *Principles of Convective Heat Transfer*. Ed. by Frederick F. Ling. 2nd ed. Mechanical Engineering Series. New York, NY: Springer New York, 2001, p. 709. ISBN: 978-1-4419-2894-8. DOI: 10.1007/978-1-4757-3488-1.

[67] Sadik Kakac, Yaman Yener, and Anchasa Pramuanjaroenkij. *Convective Heat Transfer*. 3rd ed. Boca Raton, FL, USA: CRC Press, 2013, p. 622. ISBN: 1466583444.

[68] Louis C. Burmeister. *Convective Heat Transfer*. 2nd ed. Hoboken, NJ, USA: John Wiley & Sons, Inc., 1993, p. 640. ISBN: 047157709X.

[69] Jack Philip Holman. *Heat Transfer*. Ed. by Anne Murphy, Madelaine Eichberg, and Steven Tenney. 6th ed. Singapore: McGraw-Hill Book Co., 1986.

[70] Philip Kosky, Robert Balmer, William Keat, and George Wise. *Exploring Engineering*. Elsevier, 2021. ISBN: 9780128150733. DOI: 10.1016/C2017-0-01871-2.

[71] Weiwei Li, Shuai Yang, and Atif Shamim. 'Screen printing of silver nanowires: balancing conductivity with transparency while maintaining flexibility and stretchability'. In: *npj Flexible Electronics* 3.1 (July 2019), p. 13. DOI: 10.1038/s41528-019-0057-1.

[72] Henkel. *LOCTITE ECI 1010 E&C*. Düsseldorf, 2018.

[73] Henkel. *LOCTITE EDAG PF 455B E&C*. Düsseldorf, 2014.

[74] Henkel. *LOCTITE ECI 7004HR E&C*. Düsseldorf, 2022.

[75] Henkel. *LOCTITE EDAG PM 404 E&C*. Düsseldorf, 2014.

[76] John F. Moulder, William F. Stickle, Peter E. Sobol, and Kenneth D. Bomben. *Handbook of X-ray Photoelectron Spectroscopy*. Ed. by Jill Chastain and Roger C. Jr. King. 2nd ed. Eden Prairie, Minnesota 55344: Physical Electronics, Inc., 1995. ISBN: 0-9648124-1-X.

[77] Albert Einstein. 'Über einen die Erzeugung und Verwandlung des Lichtes betreffenden heuristischen Gesichtspunkt'. In: *Annalen der Physik* 322.6 (1905), pp. 132–148. DOI: 10.1002/andp.19053220607.

[78] Dirk Hünniger. *Photoelectric Effect Schematic*. 2006. URL: https://commons.wikimedia.org/wiki/File:Photoelectric%7B%5C_%7DEffect%7B%5C_%7DSchematic-de.svg (visited on 06/05/2023).

[79] A. Carlson. *Auger Process*. 2022. URL: https://commons.wikimedia.org/wiki/File:Auger%7B%5C_%7DProcess.svg (visited on 06/05/2023).

[80] Pierre Auger. 'Sur l'effet photoélectrique composé'. In: *Journal de Physique et le Radium* 6.6 (1925), pp. 205–208. DOI: 10.1051/jphysrad:0192500606020500.

[81] Lise Meitner. 'Über die Entstehung der β-Strahl-Spektren radioaktiver Substanzen'. In: *Zeitschrift für Physik* 9.1 (Dec. 1922), pp. 131–144. DOI: 10.1007/BF01326962.

[82] Andreas Klein, Thomas Mayer, Andreas Thissen, and Wolfram Jaegermann. 'Photoelectron Spectroscopy in Materials Science and Physical Chemistry :' in: *Bunsen- Magazin* (2008), pp. 124–139.

[83] Florian Ullrich. 'Photoelectron Spectroscopic Analysis of Solution-Processed Nickel Oxide: Surface Modification and Interfaces in Organic Solar Cells'. Dissertation. Technische Universität Darmstadt, 2020.

[84] J.H. Scofield. 'Hartree-Slater subshell photoionization cross-sections at 1254 and 1487 eV'. In: *Journal of Electron Spectroscopy and Related Phenomena* 8.2 (Jan. 1976), pp. 129–137. DOI: 10.1016/0368-2048(76)80015-1.

[85] Patrick Reiser. 'Modifying and Controlling Diffusion Properties of Molecular Dopants in Organic Semiconductors'. Dissertation. Technische Universität Darmstadt, 2019.

[86] Markus Frericks. 'Organic Hole Transport Materials: Properties and Interface Formation'. Dissertation. Technische Universität Darmstadt, 2022.

[87] Vaishali Umredkar. *Automotive Sensors Future Growth*. 2020. URL: https://electronicsmaker.com/automotive-sensors-future-growtht (visited on 05/03/2023).

[88] Mary B. Alatise and Gerhard P. Hancke. 'A Review on Challenges of Autonomous Mobile Robot and Sensor Fusion Methods'. In: *IEEE Access* 8 (2020), pp. 39830–39846. DOI: 10.1109/ACCESS.2020.2975643.

[89] Ning Sun, Beibei Fan, Yahui Ding, Yuanjinsheng Liu, Yuyun Bi, Patience Afi Seglah, and Chunyu Gao. 'Analysis of the Development Status and Prospect of China's Agricultural Sensor Market under Smart Agriculture'. In: *Sensors* 23.6 (Mar. 2023), p. 3307. DOI: 10.3390/s23063307.

[90] Raghavv Raghavender Suresh, Muthaiyan Lakshmanakumar, J. B. B. Arockia Jayalatha, K. S. Rajan, Swaminathan Sethuraman, Uma Maheswari Krishnan, and John Bosco Balaguru Rayappan. 'Fabrication of screen-printed electrodes: opportunities and challenges'. In: *Journal of Materials Science* 56.15 (May 2021), pp. 8951–9006. DOI: 10.1007/s10853-020-05499-1.

[91] Rainer Bäuerle, Pariya Nazari, Johannes Zimmermann, Christian Melzer, Gerardo Hernandez-Sosa, and Wolfgang Kowalsky. 'Fully Printed PTC Based Heat Transfer Sensor Array as Liquid Level Sensor'. In: *Advanced Sensor Research* (2024). DOI: 10.1002/adsr.202400060.

[92] D. M. Bigg. 'The effect of compounding on the conductive properties of EMI shielding compounds'. In: *Advances in Polymer Technology* 4.34 (1984), pp. 255–266. DOI: 10.1002/adv.1984.060040305.

[93] Donald M. Bigg Battelle. 'Conductive polymeric compositions'. In: *Polymer Engineering and Science* 17.12 (Dec. 1977), pp. 842–847. DOI: 10.1002/pen.760171206.

[94] Shahriar Ghaffari-Mosanenzadeh, Omid Aghababaei Tafreshi, Erik Dammen-Brower, Elmira Rad, Mohammad Meysami, and Hani E. Naguib. 'A review on high thermally conductive polymeric composites'. In: *Polymer Composites* 43.2 (Feb. 2022), pp. 692–711. DOI: 10.1002/pc.26420.

[95] Masao Sumita, Kazuya Sakata, Shigeo Asai, Keizo Miyasaka, and Hideaki Nakagawa. 'Dispersion of fillers and the electrical conductivity of polymer blends filled with carbon black'. In: *Polymer Bulletin* 25.2 (1991), pp. 265–271. DOI: 10.1007/BF00310802.

[96] You Zeng, Guixia Lu, Han Wang, Jinhong Du, Zhe Ying, and Chang Liu. 'Positive temperature coefficient thermistors based on carbon nanotube/polymer composites'. In: *Scientific Reports* 4 (2014), pp. 1–7. DOI: 10.1038/srep06684.

[97] Hyun-Jung Choi, Moo Sung Kim, Damiro Ahn, Sang Young Yeo, and Sohee Lee. 'Electrical percolation threshold of carbon black in a polymer matrix and its application to antistatic fibre'. In: *Scientific Reports* 9.1 (Apr. 2019), p. 6338. DOI: 10.1038/s41598-019-42495-1.

[98] C. C. Chen and Y. C. Chou. 'Electrical-Conductivity Fluctuations near the Percolation Threshold'. In: *Physical Review Letters* 54.23 (June 1985), pp. 2529–2532. DOI: 10.1103/PhysRevLett.54.2529.

[99] Yu-Dong Shi, Jie Li, Yan-Jun Tan, Yi-Fu Chen, and Ming Wang. 'Percolation behavior of electromagnetic interference shielding in polymer/multi-walled carbon nanotube nanocomposites'. In: *Composites Science and Technology* 170.November 2018 (Jan. 2019), pp. 70–76. DOI: 10.1016/j.compscitech.2018.11.033.

Bibliography

[100] Long Chen and Jianming Zhang. 'Designs of conductive polymer composites with exceptional reproducibility of positive temperature coefficient effect: A review'. In: *Journal of Applied Polymer Science* 138.3 (Jan. 2021), p. 49677. DOI: 10.1002/app.49677.

[101] Charles Trudeau, Patrick Beaupré, Martin Bolduc, and Sylvain G. Cloutier. 'All inkjet-printed perovskite-based bolometers'. In: *npj Flexible Electronics* 4.1 (Dec. 2020), p. 34. DOI: 10.1038/s41528-020-00097-2.

[102] Soo-jin Park, Min-kang Seo, and Jae-rock Lee. 'PTC / NTC Behaviors of Nanostructured Carbon Black-filled HDPE Polymer Composites'. In: *Science* 2.3 (2001), pp. 159–164.

[103] Zheng Chen, Po-Chun Hsu, Jeffrey Lopez, Yuzhang Li, John W. F. To, Nan Liu, Chao Wang, Sean C. Andrews, Jia Liu, Yi Cui, and Zhenan Bao. 'Fast and reversible thermoresponsive polymer switching materials for safer batteries'. In: *Nature Energy* 1.1 (Jan. 2016), p. 15009. DOI: 10.1038/nenergy.2015.9.

[104] Kunmo Chu, Sung-Chul Lee, Sangeui Lee, Dongearn Kim, Changyoul Moon, and Sung-Hoon Park. 'Smart conducting polymer composites having zero temperature coefficient of resistance'. In: *Nanoscale* 7.2 (2015), pp. 471–478. DOI: 10.1039/C4NR04489D.

[105] Hua Deng, Tetyana Skipa, Rui Zhang, Dirk Lellinger, Emiliano Bilotti, Ingo Alig, and Ton Peijs. 'Effect of melting and crystallization on the conductive network in conductive polymer composites'. In: *Polymer* 50.15 (July 2009), pp. 3747–3754. DOI: 10.1016/j.polymer.2009.05.016.

[106] Imran Ali, Alexandr Shchegolkov, Aleksei Shchegolkov, Natalya Zemtsova, Vladimir Bogoslovskiy, Gulnara Shigabaeva, Evgeny Galunin, Iqbal Hussain, Abdulraheem S. A. Almalki, Meshari A. Alsharif, and Mohammed Issa Alahmdi. 'Preparation and application practice of temperature self-regulating flexible polymer electric heaters'. In: *Polymer Engineering & Science* 62.3 (Mar. 2022), pp. 730–742. DOI: 10.1002/pen.25880.

[107] Microchip. *ATmega640/V-1280/V-1281/V-2560/V-2561/V*. Chandler, Arizona, USA: Microchip, 2020.

[108] Zhe Yang, Hongdan Peng, Weizhi Wang, and Tianxi Liu. 'Crystallization behavior of poly(ϵ-caprolactone)/layered double hydroxide nanocomposites'. In: *Journal of Applied Polymer Science* 116.5 (2010), pp. 2658–2667. DOI: 10.1002/app.

[109] Dan Yang, Xinxin Kong, Yufeng Ni, Dahai Gao, Bin Yang, Yangyang Zhu, and Liqun Zhang. 'Novel nitrile-butadiene rubber composites with enhanced thermal conductivity and high dielectric constant'. In: *Composites Part A: Applied Science and Manufacturing* 124.May (Sept. 2019), p. 105447. DOI: 10.1016/j.compositesa.2019.05.015.

[110] V. S. Vinod, Siby Varghese, and Baby Kuriakose. 'Aluminum powder filled nitrile rubber composites'. In: *Journal of Applied Polymer Science* 91.5 (Mar. 2004), pp. 3156–3161. DOI: 10.1002/app.13472.

[111] K. Kadoya, N. Matsunaga, and A. Nagashima. 'Viscosity and Thermal Conductivity of Dry Air in the Gaseous Phase'. In: *Journal of Physical and Chemical Reference Data* 14.4 (Oct. 1985), pp. 947–970. DOI: 10.1063/1.555744.

[112] Gonglun Chen and Daniel Tao. 'An experimental study of stability of oil–water emulsion'. In: *Fuel Processing Technology* 86.5 (Feb. 2005), pp. 499–508. DOI: 10.1016/j.fuproc.2004.03.010.

[113] S. Mataumoto and W.W. Kang. 'Formation and Applications of Multiple Emulsions'. In: *Journal of Dispersion Science and Technology* 10.4-5 (Jan. 1989), pp. 455–482. DOI: 10.1080/01932698908943184.

[114] Mahmoud Meribout, Ahmed Al, and Khamis Al. 'Interface Layers Detection in Oil Field Tanks: A Critical Review'. In: *Expert Systems for Human, Materials and Automation*. InTech, Oct. 2011. DOI: 10.5772/17166.

[115] Ramon Casanella, Óscar Casas, and Ramon Pallàs-Areny. 'Oil-Water Interface Level Sensor Based on an Electrode Array'. In: *2006 IEEE Instrumentation and Measurement Technology Conference Proceedings*. April. IEEE, 2006, pp. 710–713. ISBN: 0-7803-9360-0. DOI: 10.1109/IMTC.2006.235823.

[116] W. Q. Yang, M. R. Brant, and M. S. Beck. 'A multi-interface level measurement system using a segmented capacitance sensor for oil separators'. In: *Measurement Science and Technology* 5.9 (1994), pp. 1177–1180. DOI: 10.1088/0957-0233/5/9/021.

[117] Daniel Paczesny, Grzegorz Tarapata, Marzęcki Michał, and Ryszard Jachowicz. 'The Capacitive Sensor for Liquid Level Measurement Made with Ink-jet Printing Technology'. In: *Procedia Engineering* 120 (2015), pp. 731–735. DOI: 10.1016/j.proeng.2015.08.776.

[118] Binfeng Yun, Na Chen, and Yiping Cui. 'Highly sensitive liquid-level sensor based on etched fiber bragg grating'. In: *IEEE Photonics Technology Letters* 19.21 (2007), pp. 1747–1749. DOI: 10.1109/LPT.2007.905093.

Bibliography

[119] Qi Liu, Tao Liu, Tao He, Hao Li, Zhijun Yan, Lin Zhang, and Qizhen Sun. 'High resolution and large sensing range liquid level measurement using phase-sensitive optic distributed sensor'. In: *Optics Express* 29.8 (Apr. 2021), p. 11538. DOI: 10.1364/OE.412935.

[120] S. Mann, S. Lindner, F. Barbon, S. Linz, A. Talai, R. Weigel, and A. Koelpin. 'A tank level sensor based on Six-Port technique comprising a quasi-TEM waveguide'. In: *2014 IEEE Topical Conference on Wireless Sensors and Sensor Networks (WiSNet)*. 1. IEEE, Jan. 2014, pp. 4–6. ISBN: 978-1-4799-2300-7. DOI: 10.1109/WiSNet.2014.6825512.

[121] Minoru Takeda, Yu Matsuno, Itaru Kodama, Hiroaki Kumakura, and Chikara Kazama. 'Application of MgB2 Wire to Liquid Hydrogen Level Sensor—External-Heating-Type MgB2 Level Sensor'. In: *IEEE Transactions on Applied Superconductivity* 19.3 (June 2009), pp. 764–767. DOI: 10.1109/TASC.2009.2019033.

[122] Jonggan Hong, Young Soo Chang, and Dongsik Kim. 'Development of a micro liquid-level sensor for harsh environments using a periodic heating technique'. In: *Measurement Science and Technology* 21.10 (Oct. 2010), p. 105408. DOI: 10.1088/0957-0233/21/10/105408.

[123] Bernd D. Zimmermann, Tung-Sheng Yang, and Thomas C. Anderson. *Liquid level sensor using thick film substrate*. 2009.

[124] F A Khan, A Yousaf, and L M Reindl. 'Build-up detection and level monitoring by using capacitive glocal technique'. In: *2016 European Frequency and Time Forum (EFTF)*. IEEE, Apr. 2016, pp. 1–4. ISBN: 978-1-5090-0720-2. DOI: 10.1109/EFTF.2016.7477836.

[125] Seokgyu Ryu, Kiho Kim, and Jooheon Kim. 'Silane surface treatment of boron nitride to improve the thermal conductivity of polyethylene naphthalate requiring high temperature molding'. In: *Polymer Composites* 39.S3 (June 2018), E1692–E1700. DOI: 10.1002/pc.24680.

[126] Lin Li and D. D. L. Chung. 'Thermally conducting polymer-matrix composites containing both AIN particles and SiC whiskers'. In: *Journal of Electronic Materials* 23.6 (June 1994), pp. 557–564. DOI: 10.1007/BF02670659.

[127] Joseph Hilsenrath, Charles W. Beckett, William S. Benedict, Lilla Fano, Harold J. Hoge, Joseph F. Masi, Ralph L. Nuttall, Yeram S. Touloukian, and Harold W. Woolley. *Circular of the Bureau of Standards no. 564: tables of thermal properties of gases comprising tables of thermodynamic and transport properties of air, argon, carbon dioxide, carbon monoxide hydrogen, nitrogen, oxygen, and steam*. Gaithersburg, MD, USA: Commerce Department, National Institute of Standards and Technology (NIST), 1955.

[128] Schott AG. *BOROFLOAT 33 – Technical Data*. Jena, 2019.

[129] Oscar Kenneth Bates. 'Thermal Conductivity of Liquids'. In: *Industrial & Engineering Chemistry* 25.4 (Apr. 1933), pp. 431–437. DOI: 10.1021/ie50280a021.

[130] 'Appendix'. In: *The Extra-Virgin Olive Oil Handbook*. Ed. by Claudio Peri. Chichester, UK: John Wiley & Sons, Ltd, Mar. 2014, pp. 349–360. DOI: 10.1002/9781118460412.app1.

[131] Umesh Gaur and Bernhard Wunderlich. 'Heat Capacity and Other Thermodynamic Properties of Linear Macromolecules. V. Polystyrene'. In: *Journal of Physical and Chemical Reference Data* 11.2 (Apr. 1982), pp. 313–325. DOI: 10.1063/1.555663.

[132] Pedro Fierro, ed. *The Water Encyclopedia, Third Edition*. CRC Press, Feb. 2007. ISBN: 978-1-56670-645-2. DOI: 10.1201/9781420012583.

[133] Hussein K. Abdel-Aal, Mohamed A. Aggour, and Mohamed A. Fahim. *Petroleum and Gas Field Processing*. 1st Editio. Boca Raton: CRC Press, July 2003, p. 364. ISBN: 9780429258497. DOI: 10.1201/9780429258497.

[134] Peng-Cheng Chen and Zhi-Kang Xu. 'Mineral-Coated Polymer Membranes with Superhydrophilicity and Underwater Superoleophobicity for Effective Oil/Water Separation'. In: *Scientific Reports* 3.1 (Sept. 2013), p. 2776. DOI: 10.1038/srep02776.

[135] Turlough F. Guerin. 'Heavy equipment maintenance wastes and environmental management in the mining industry'. In: *Journal of Environmental Management* 66.2 (Oct. 2002), pp. 185–199. DOI: 10.1006/jema.2002.0583.

[136] Kuo-Yann Lai. *Liquid Detergents*. 2nd ed. CRC Press, 2005. ISBN: 978-1-4200-2790-7.

[137] RS Components. *Non-Silicone Thermal Grease , 4W/m · K*. 2022.

[138] Christoph Clauser and Ernst Huenges. 'Thermal Conductivity of Rocks and Minerals'. In: *Rock Physics & Phase Relations: A Handbook of Physical Constants*. Ed. by Thomas J. Ahrens. Vol. 3. American Geophysical Union, 1995, pp. 105–126. DOI: 10.1029/RF003p0105.

[139] 3M. *Transferklebebänder ohne Träger Serie 200MP*. Neuss, 2006.

[140] Zhihua Zhou, Dichen Li, Junhua Zeng, and Zhengyu Zhang. 'Rapid fabrication of metal-coated composite stereolithography parts'. In: *Proceedings of the Institution of Mechanical Engineers, Part B: Journal of Engineering Manufacture* 221.9 (Sept. 2007), pp. 1431–1440. DOI: 10.1243/09544054JEM827.

[141] Zhihua Zhou, Dichen Li, Zhengyu Zhang, and Junhua Zeng. 'Design and fabrication of a hybrid surface-pressure airfoil model based on rapid prototyping'. In: *Rapid Prototyping Journal* 14.1 (Jan. 2008), pp. 57–66. DOI: 10.1108/13552540810841571.

[142] Gary E. Erickson. 'High Angle-of-Attack Aerodynamics'. In: *Annual Review of Fluid Mechanics* 27.1 (Jan. 1995), pp. 45–88. DOI: `10.1146/annurev.fl.27.010195.000401`.

[143] Lindsay J. Lina and Martin T. Moul. 'A simulator study of T-tail aircraft in deep stall conditions'. In: *Aircraft Design and Technology Meeting*. Reston, Virigina: American Institute of Aeronautics and Astronautics, Nov. 1965. DOI: `10.2514/6.1965-781`.

[144] K. G. Gousman, J. C. Juang, R. C. Loschke, and R. H. Rooney. 'Aircraft deep stall analysis and recovery'. In: *18th Atmospheric Flight Mechanics Conference*. American Institute of Aeronautics and Astronautics, Aug. 1991. DOI: `10.2514/6.1991-2888`.

[145] Obi I. Iloputaife. 'Design of Deep Stall Protection for the C-17A'. In: *Journal of Guidance, Control, and Dynamics* 20.4 (July 1997), pp. 760–767. DOI: `10.2514/2.4109`.

[146] Duc H. Nguyen, Mark H. Lowenberg, and Simon A. Neild. 'Analysing dynamic deep stall recovery using a nonlinear frequency approach'. In: *Nonlinear Dynamics* 108.2 (Apr. 2022), pp. 1179–1196. DOI: `10.1007/s11071-022-07283-z`.

[147] A. Van Brecht, H. Hens, J-L. Lemaire, J.M. Aerts, P. Degraeve, and D. Berckmans. 'Quantification of the heat exchange of chicken eggs'. In: *Poultry Science* 84.3 (Mar. 2005), pp. 353–361. DOI: `10.1093/ps/84.3.353`.

[148] Avraham Shitzer. 'Wind-chill-equivalent temperatures: regarding the impact due to the variability of the environmental convective heat transfer coefficient'. In: *International Journal of Biometeorology* 50.4 (Mar. 2006), pp. 224–232. DOI: `10.1007/s00484-005-0011-x`.

[149] Dongwei Yu, Dayong Zhang, Lin Wu, Xiangyi Kong, and Qianjin Yue. 'Analysis of the Influence of Convection Heat Transfer in Circular Tubes on Ships in a Polar Environment'. In: *Atmosphere* 13.2 (2022), pp. 1–10. DOI: `10.3390/atmos13020149`.

[150] Chaima Mahjoubi, Jean Christophe Olivier, Sondes Skander-mustapha, Mohamed Machmoum, and Ilhem Slama-belkhodja. 'An improved thermal control of open cathode proton exchange membrane fuel cell'. In: *International Journal of Hydrogen Energy* 44.22 (2019), pp. 11332–11345. DOI: `10.1016/j.ijhydene.2018.11.055`.

[151] Hw-lab. *Noctua NF-S12B FLX — designed to be quiet.* URL: `https://hw-lab.com/noctua-nf-s12b-flx-review.html/3` (visited on 11/18/2022).

[152] Noctua. *NF-R8 redux-1200.* URL: `https://noctua.at/en/nf-r8-redux-1200/specification` (visited on 11/28/2022).

[153] Weatherspark.com. *Klima und durchschnittliches Wetter das ganze Jahr über in Heidelberg*. URL: https://de.weatherspark.com/y/60978/Durchschnittswetter-in-Heidelberg-Deutschland-das-ganze-Jahr-%7B%5C%22%7Bu%7D%7Dber (visited on 04/19/2021).

[154] Tyler Sonsalla, Arden L. Moore, W.J. Meng, Adarsh D. Radadia, and Leland Weiss. '3-D printer settings effects on the thermal conductivity of acrylonitrile butadiene styrene (ABS)'. In: *Polymer Testing* 70.May (Sept. 2018), pp. 389–395. DOI: 10.1016/j.polymertesting.2018.07.018.

[155] J. D. Macias, J. Bante-Guerra, F. Cervantes-Alvarez, G. Rodrìguez-Gattorno, O. Arés-Muzio, H. Romero-Paredes, C. A. Arancibia-Bulnes, V. Ramos-Sánchez, H. I. Villafán-Vidales, J. Ordonez-Miranda, R. Li Voti, and J. J. Alvarado-Gil. 'Thermal Characterization of Carbon Fiber-Reinforced Carbon Composites'. In: *Applied Composite Materials* 26.1 (Feb. 2019), pp. 321–337. DOI: 10.1007/s10443-018-9694-0.

[156] A. Tropis, M. Thomas, J. L. Bounie, and P. Lafon. 'Certification of the Composite Outer Wing of the ATR72'. In: *Proceedings of the Institution of Mechanical Engineers, Part G: Journal of Aerospace Engineering* 209.4 (Dec. 1995), pp. 327–339. DOI: 10.1243/PIME_PROC_1995_209_307_02.

[157] Jian-Qiang Zhong, Mengen Wang, William H. Hoffmann, Matthijs A. van Spronsen, Deyu Lu, and J. Anibal Boscoboinik. 'Synchrotron-based ambient pressure X-ray photoelectron spectroscopy of hydrogen and helium'. In: *Applied Physics Letters* 112.9 (Feb. 2018), p. 091602. DOI: 10.1063/1.5022479.

[158] B J Lindberg, K Hamrin, G Johansson, U Gelius, A Fahlman, C Nordling, and K Siegbahn. 'Molecular Spectroscopy by Means of ESCA II. Sulfur compounds. Correlation of electron binding energy with structure'. In: *Physica Scripta* 1.5-6 (May 1970), pp. 286–298. DOI: 10.1088/0031-8949/1/5-6/020.

[159] Jin-Hao Jhang, J. Anibal Boscoboinik, and Eric I. Altman. 'Ambient pressure x-ray photoelectron spectroscopy study of water formation and adsorption under two-dimensional silica and aluminosilicate layers on Pd(111)'. In: *The Journal of Chemical Physics* 152.8 (Feb. 2020), p. 084705. DOI: 10.1063/1.5142621.

[160] Cao-Thang Dinh, Ankit Jain, F. Pelayo García de Arquer, Phil De Luna, Jun Li, Ning Wang, Xueli Zheng, Jun Cai, Benjamin Z. Gregory, Oleksandr Voznyy, Bo Zhang, Min Liu, David Sinton, Ethan J. Crumlin, and Edward H. Sargent. 'Multi-site electrocatalysts for hydrogen evolution in neutral media by destabilization of water molecules'. In: *Nature Energy* 4.2 (Dec. 2018), pp. 107–114. DOI: 10.1038/s41560-018-0296-8.

[161] Emily G. Bittle, James I. Basham, Thomas N. Jackson, Oana D. Jurchescu, and David J. Gundlach. 'Mobility overestimation due to gated contacts in organic field-effect transistors'. In: *Nature Communications* 7.1 (Mar. 2016), p. 10908. DOI: 10.1038/ncomms10908.

[162] Iain McCulloch, Alberto Salleo, and Michael Chabinyc. 'Avoid the kinks when measuring mobility'. In: *Science* 352.6293 (June 2016), pp. 1521–1522. DOI: 10.1126/science.aaf9062.

[163] Takafumi Uemura, Cedric Rolin, Tung-Huei Ke, Pavlo Fesenko, Jan Genoe, Paul Heremans, and Jun Takeya. 'On the Extraction of Charge Carrier Mobility in High-Mobility Organic Transistors'. In: *Advanced Materials* 28.1 (Jan. 2016), pp. 151–155. DOI: 10.1002/adma.201503133.

[164] Hagen Klauk. 'Will We See Gigahertz Organic Transistors?' In: *Advanced Electronic Materials* 4.10 (Oct. 2018), p. 1700474. DOI: 10.1002/aelm.201700474.

[165] Martin Weis, Jack Lin, Dai Taguchi, Takaaki Manaka, and Mitsumasa Iwamoto. 'Insight into the contact resistance problem by direct probing of the potential drop in organic field-effect transistors'. In: *Applied Physics Letters* 97.26 (Dec. 2010), p. 263304. DOI: 10.1063/1.3533020.

[166] Stephen Forrest, Paul Burrows, and Mark Thompson. 'The dawn of organic electronics'. In: *IEEE Spectrum* 37.8 (Aug. 2000), pp. 29–34. DOI: 10.1109/6.861775.

[167] Henning Sirringhaus. 'Reliability of Organic Field-Effect Transistors'. In: *Advanced Materials* 21 (Oct. 2009), pp. 3859–3873. DOI: 10.1002/adma.200901136.

[168] Peter A. Bobbert, Abhinav Sharma, Simon G. J. Mathijssen, Martijn Kemerink, and Dago M. de Leeuw. 'Operational Stability of Organic Field-Effect Transistors'. In: *Advanced Materials* 24.9 (Mar. 2012), pp. 1146–1158. DOI: 10.1002/adma.201104580.

[169] Henning Sirringhaus. '25th Anniversary Article: Organic Field-Effect Transistors: The Path Beyond Amorphous Silicon'. In: *Advanced Materials* 26.9 (Mar. 2014), pp. 1319–1335. DOI: 10.1002/adma.201304346.

Acronyms

ABS	acrylonitrile butadiene styrene
ADC	analog-to-digital converter
AI	artificial intelligence
CL	current limiter
CPC	conductive polymer composite
CTS	computer to screen
CV	constant voltage
HOMO	highest occupied molecular orbital
IPA	isopropyl alcohol
I-V curve	current-voltage curve
LUMO	lowest unoccupied molecular orbital
OFET	organic field effect transistor
PEN	polyethylene naphtalate
PES	photoelectron spectroscopy
PET	polyethylene terephthalate
PI	polyimide
PP	polypropylene
PSU	power supply unit
PTC	positive temperature coefficient (of resistance)
SEC	secondary electron cutoff
Si	silicon
SMU	source measure unit
SR	shunt resistor
TDS	technical data sheet
TLM	transmission line method
UHV	ultra-high vacuum
UPS	ultraviolet photoelectron spectroscopy
XPS	X-ray photoelectron spectroscopy

List of Figures

List of Tables

Publications, Presentations, Theses

Journal Publications

1. Patrick Reiser, Frank S. Benneckendorf, Marc-Michael Barf, Lars Müller, **Rainer Bäuerle**, Sabina Hillebrandt, Sebastian Beck, Robert Lovrincic, Eric Mankel, Jan Freudenberg, Daniel Jänsch, Wolfgang Kowalsky, Annemarie Pucci, Wolfram Jaeger-mann, Uwe H.F. Bunz, and Klaus Müllen. 'n-Type Doping of Organic Semiconductors: Immobilization via Covalent Anchoring'. In: *Chemistry of Materials* 31.11 (June 2019), pp. 4213–4221. DOI: 10.1021/acs.chemmater.9b01150. URL: https://pubs.acs.org/doi/10.1021/acs.chemmater.9b01150

2. Ali Abdulkarim, Karl-Philipp Strunk, **Rainer Bäuerle**, Sebastian Beck, Hanna Makowska, Tomasz Marszalek, Annemarie Pucci, Christian Melzer, Daniel Jänsch, Jan Freudenberg, Uwe H.F. Bunz, and Klaus Müllen. 'Small change, big impact: The shape of precursor polymers governs poly-p-phenylene synthesis'. In: *Macromolecules* 52.12 (2019), pp. 4458–4463. DOI: 10.1021/acs.macromol.9b00792

3. Masoud Payandeh, Vahid Ahmadi, Farzaneh Arabpour Roghabadi, Pariya Nazari, Fatemeh Ansari, Philipp Brenner, **Rainer Bäuerle**, Marius Jakoby, Uli Lemmer, Ian A. Howard, Bryce S. Richards, Ulrich W. Paetzold, and Bahram Abdollahi Nejand. 'High-Brightness Perovskite Light-Emitting Diodes Using a Printable Silver Microflake Contact'. In: *ACS Applied Materials and Interfaces* 12.10 (2020), pp. 11428–11437. DOI: 10.1021/acsami.9b18527

4. Ali Abdulkarim, Marvin Nathusius, **Rainer Bäuerle**, Karl-Philipp Strunk, Sebastian Beck, Hans Joachim Räder, Annemarie Pucci, Christian Melzer, Daniel Jänsch, Jan Freudenberg, Uwe H. F. Bunz, and Klaus Müllen. 'Beyond p-Hexaphenylenes: Synthesis of Unsubstituted p-Nonaphenylene by a Precursor Protocol'. In: *Chemistry – A European Journal* (2020). DOI: 10.1002/chem.202001531

5. Marc-Michael Barf, Frank S. Benneckendorf, Patrick Reiser, **Rainer Bäuerle**, Wolfgang Köntges, Lars Müller, Martin Pfannmöller, Sebastian Beck, Eric Mankel, Jan Freudenberg, Daniel Jänsch, Jean-Nicolas Tisserant, Robert Lovrincic, Rasmus R. Schröder, Uwe H. F. Bunz, Annemarie Pucci, Wolfram Jaegermann, Wolfgang Kowalsky, and Klaus Müllen. 'Compensation of Oxygen Doping in p-Type Organic Field-Effect Transistors Utilizing Immobilized n-Dopants'. In: *Advanced Materials Technologies* 2000556 (Nov. 2020), p. 2000556. DOI: 10.1002/admt.202000556.

URL: https://onlinelibrary.wiley.com/doi/10.1002/admt.202000556

6. Xiaokun Huang, **Rainer Bäuerle**, Felix Scherz, Jean-Nicolas Tisserant, Wolfgang Kowalsky, Robert Lovrinčić, and Gerardo Hernandez-Sosa. 'Improved performance of perovskite light-emitting diodes with a NaCl doped PEDOT:PSS hole transport layer'. In: *Journal of Materials Chemistry C* 9.12 (2021), pp. 4344–4350. DOI: 10.1039/D0TC06058E. URL: http://xlink.rsc.org/?DOI=D0TC06058E

7. Xiaokun Huang, **Rainer Bäuerle**, Jean-Nicolas Tisserant, Wolfgang Kowalsky, Robert Lovrinčić, and Gerardo Hernandez-Sosa. 'Blue-emission tuning of perovskite light-emitting diodes with a simple TPBi surface treatment'. In: *MRS Communications* 11.6 (Dec. 2021), pp. 856–861. DOI: 10.1557/s43579-021-00108-x. URL: https://link.springer.com/10.1557/s43579-021-00108-x

8. N. Maximilian Bojanowski, Christian Huck, Lisa Veith, Karl-Philipp Strunk, **Rainer Bäuerle**, Christian Melzer, Jan Freudenberg, Irene Wacker, Rasmus R. Schröder, Petra Tegeder, and Uwe H. F. Bunz. 'Electron-beam lithography of cinnamate polythiophene films: conductive nanorods for electronic applications'. In: *Chemical Science* 13.26 (2022), pp. 7880–7885. DOI: 10.1039/D2SC01867E. URL: http://xlink.rsc.org/?DOI=D2SC01867E

9. Ahmed Farag, Thomas Feeney, Ihteaz M Hossain, Fabian Schackmar, Paul Fassl, Kathrin Küster, **Rainer Bäuerle**, Marco A. Ruiz-Preciado, Mario Hentschel, David B Ritzer, Alexander Diercks, Yang Li, Bahram Abdollahi Nejand, Felix Laufer, Roja Singh, Ulrich Starke, and Ulrich W Paetzold. 'Evaporated Self-Assembled Monolayer Hole Transport Layers: Lossless Interfaces in p-i-n Perovskite Solar Cells'. In: *Advanced Energy Materials* (Jan. 2023), p. 2203982. DOI: 10.1002/aenm.202203982. URL: https://onlinelibrary.wiley.com/doi/10.1002/aenm.202203982

10. Pariya Nazari, **Rainer Bäuerle**, Johannes Zimmermann, Christian Melzer, Christopher Schwab, Anna Smith, Wolfgang Kowalsky, Jasmin Aghassi-Hagmann, Gerardo Hernandez-Sosa, and Uli Lemmer. 'Piezoresistive Free-standing Microfiber Strain Sensor for High-resolution Battery Thickness Monitoring'. In: *Advanced Materials* (Mar. 2023), p. 2212189. DOI: 10.1002/adma.202212189. URL: https://onlinelibrary.wiley.com/doi/10.1002/adma.202212189

11. **Rainer Bäuerle**, Pariya Nazari, Johannes Zimmermann, Christian Melzer, Gerardo Hernandez-Sosa, and Wolfgang Kowalsky. 'Fully Printed PTC Based Heat Transfer Sensor Array as Liquid Level Sensor'. In: *Advanced Sensor Research* (2024). DOI: 10.1002/adsr.202400060. URL: https://onlinelibrary.wiley.com/doi/10.1002/adsr.202400060

Conference Presentations

1. Rainer Bäuerle, Christian Willig, Pariya Nazari, Jean-Nicolas Tisserant, Johannes Zimmermann, Christian Melzer, Wolfgang Kowalsky, and Uwe Bunz. *Highly Sensitive, Fully Screen-Printed Sensor Matrix Based on a PTC Material for Sensing Thermal Energy Flow.* Talk, MRS Spring 2022, virtual.

Co-supervised Theses

1. Melina Maag, Heidelberg University: *Natrium-Ionen sensitive Sensoren und Leitfähigkeit von Silbernanodraht-Schichten*, research internship.

Acknowledgment

I would like to express my thanks to everyone who contributed to this work in any matter. My thanks goes in particular to...

... Prof Wolfgang Kowalsky for the possibility of this work, his supervision of this thesis, and the freedom to shape this work in my direction.

... Prof Erwin Peiner for being the second referee of my thesis and the head of committee Prof Jörg Schöbel.

... Prof Uwe Bunz for the employment and financial support of this work and Kerstin Windisch for support in many administrative matters.

... PD Christian Melzer and Dr Johannes Zimmermann for supervision, advice and support of this work. Without the many fruitful discussions, this work would have been much more difficult.

... Dr Karl-Philipp Strunk and Dr Stefan Schlisske for collaboration with the OFETs.

... Benjamin Obermayer for screen printing my samples with me.

... Dr Kai Exner and Sebastian Raths for supplying the organic semiconductor and partnering in characterizing material and OFETs.

... Pariya Nazari and Christian Willig for being PhD candidates in the same project and many nice and helpful discussion.

... the employees at iL, both of universities and companies. My thanks go to Prof Gerardo Hernandez-Sosa for his scientific support towards the end as well as including my socially into his group where I was warmly welcome by everyone. I also thank the former analytics at iL, especially Markus Frericks for introducing the PES system to me.

... Conrad, Peter, and Rebekka for proof reading my thesis.

... my family who always supported me.

This list is by far not complete. I cannot sufficiently express my gratitude to all of you. Thank you!